GENETICS OF HOST-PARASITE INTERACTION

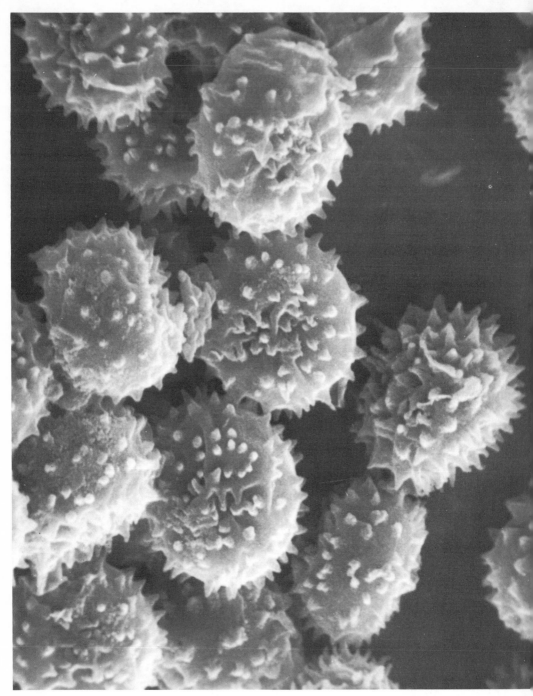

Teliospores of *Ustilago maydis* seen in scanning electron microscope. Spores were coated with carbon and gold-palladium alloy ($\times$ 7,000) (U.C. Banerjee: unpublished).

GENETICS OF
HOST-PARASITE INTERACTION

Peter R. Day
THE CONNECTICUT AGRICULTURAL EXPERIMENT STATION

W. H. FREEMAN AND COMPANY
San Francisco

A SERIES OF BOOKS IN THE BIOLOGY OF PLANT PATHOGENS

EDITORS:
Arthur Kelman
Luis Sequeira

Cover: Advancing mycelium of the powdery mildew fungus (*Erysiphe graminis hordei*) on barley seen in a scanning electron microscope × 1000. Photograph from Day and Scott (1973). Prepared by John Hardy, University of Queensland.

Library of Congress Cataloging in Publication Data

Day, Peter R   1928–
    Genetics of host-parasite interaction.

    Bibliography: p.
    1.   Plant genetics.   2.   Fungal genetics.   3.   Host— parasite relationships.   I.   Title.
QH433.D39   581.2'32   73-17054
ISBN 0-7167-0844-2

Printed in the United States of America

1   2   3   4   5   6   7   8   9

To Sue, Catherine, Rupert, and Bill

CONTENTS

PREFACE

Had breeding for resistance been a universally successful and reliable means of controlling plant parasites, there would be no reason for this book. But parasites vary and many have circumvented host-plant resistance. There are many papers describing such variation in parasites and even more on breeding for resistance in crop plants, but there are very few general accounts which deal with the genetics of both plant hosts and parasites and their interactions. I have tried to remedy this deficiency. Although my discussion is based largely on fungal parasites, I also draw on insects, nematodes, bacteria, and viruses. My chief concern is the work of the last ten years, although I have included some historical perspectives. My approach is speculative. This is especially true of Chapter 5, which explores gene function, an area in which our knowledge is surprisingly slim and where conceptual frameworks help to focus ideas. I have written for an audience of advanced undergraduates, research students, and researchers in genetics, plant breeding, plant pathology, entomology, and related fields. I have assumed a basic knowledge of genetics in my reader, but since information on the parasites is scattered, I have tried to introduce the nonspecialist to each major group that I treat. Important developments in the study of plant-parasite interaction include the extension of Flor's gene-for-gene concept to a range of interactions, the use of isogenic lines and temperature-sensitive genes in the study of the biochemistry of disease resistance, and the heightened interest in biological controls arising from dissatisfaction with pesticides. These and other topics are discussed.

Wise and intelligent methods of crop protection can only be based on an appreciation of the principles of host-parasite interaction and the consequences of interfering with it. Entomologists, plant pathologists, and plant breeders have learned much from each other, but will do so more readily with a clearer understanding of the genetic basis of their problems.

I am indebted to many colleagues and friends. The following provided copies of unpublished manuscripts: Drs. J. R. Aist and P. H. Williams, D. M. Boone, J. M. Daly, A. H. Ellingboe, G. Fleischman, S. G. Georgopoulos, M. Heath, A. L. Hooker and K. M. S. Saxena, J. Kuć, F. M. Latterell, G. M. E. Mayo and K. W. Shepherd, C. Person and G. Sidhu, R. A. Robinson, R. A. Rohde, D. S. Shaw, W. Sitterly, and L. Slootmaker. The following provided photographs: Drs. J. R. Aist and P. H. Williams, E. P. Baker and R. A. McIntosh, U. C. Banerjee, D. M. Boone, R. L. Gallun, M. and I. Heath, H. Knutson, L. J. Littlefield and C. E. Bracker, Mrs. K. Maeda Nishimoto and C. E. Bracker and G. Zentmyer and D. C. Erwin.

The published sources of these Figures were 3.1, *Journal of Cell Biology;* 3.2, *Phytopathology;* 3.5, 3.6: A,C,D,E, *American Journal of Botany;* 3.6: B, *American Naturalist;* 4.2: A,B, Kansas Agricultural Experiment Station C,D, Indiana Agricultural Experiment Station; 3.3, 5.2, *Physiological Plant Pathology;* 5.3, *Protoplasma;* and 5.6: A, *Proceedings of the Linnean Society of New South Wales.*

I am especially indebted to my colleagues at The Connecticut Agricultural Experiment Station—Mrs. Sandra Anagnostakis, Dr. Raimon Beard, Dr. Carl Clayberg, Dr. Kenneth Hanson, Dr. James Horsfall, Dr. James Kring, Dr. David Sands, Dr. Paul Waggoner, and Dr. Israel Zelitch—for reading and criticizing parts of my manuscript. Dr. James Aist, Dr. Mike Daly, Dr. Donald Erwin, Dr. Robert Gallun, and Dr. Kenneth Scott were also kind enough to comment on some sections. I am also grateful to Dr. Mannon Gallegly, Dr. Arthur Kelman, and Dr. Luis Sequeira for their comments and criticisms of the entire text. The errors and misinterpretations that may remain are my responsibility. I thank Miss Carolyn Staehly and Mrs. Evelyn Breuler for their help in preparing the manuscript.

My final thanks go to the Director and Board of Control of The Connecticut Agricultural Experiment Station for their support and encouragement.

April 1973   *Peter R. Day*

GENETICS OF HOST-PARASITE INTERACTION

1

THE HOST-PARASITE INTERACTION

Each host-parasite interaction is a struggle for survival between two organisms. The host plant, already competing with other plants for space, light, and food materials, has the added burden of supporting another organism. The parasite not only usurps its host's food but also actively impairs its ability to make more food by invading and destroying host tissue. The consequences for the host may be trivial, or if it dies before reproduction, may threaten its survival as a species. The parasite must produce many offspring so that at least one may encounter a fresh host at an opportune time to ensure survival. The parasite must resist or avoid the defense mechanisms which ward off the myriads of other organisms with which the host comes in contact. If the parasite is specialized so that it will reproduce only on certain host species, it must not be so destructive that it eliminates the host individuals entirely or it will die out along with them.

In nature a state of balance exists, and surviving plants and their parasites are capable of coexistence. If parasites reduce both inter- and intraspecific host competition, then they may have some adaptive value to their hosts. However, much that we shall learn of their genetic flexibility suggests that during evolution parasites have been kept in check by the requirement to conserve their hosts for their future survival.

The genetic basis of this plant-parasite interaction is the subject of this book. Most of our knowledge of this interaction arose as a result of a need to cope with the consequences of our altering the balance between host and parasite by growing crops for food and fiber. Large-scale culture of genetically uniform crops is unnatural and leads, inevitably, to the possibility of equally large-scale crop destruction by parasites. To keep the balance in our favor, we developed the sciences of plant pathology and entomology to understand parasites better, know their limitations, and be able to contain and control them. Breeding for resistance became a powerful tool to protect crops, but it revealed a corresponding ability to fight back on the part of the parasites. The resulting perturbations and fluctuations, which were only minor on an evolutionary scale, caused major epidemics and much human suffering. Man prepared his crops in a way that seemed almost to invite their mass destruction. Knowledge of parasites and parasitism, still far from complete, has become much clearer in the last 30 years. Pest control is still far from satisfactory, but without it life as we know it today would be very different. The increasing demands made on agriculture as the population of the world continues to grow make it imperative to find better, and ecologically sound, methods of protecting crops.

THE ROLE OF GENETICS

All biological phenomena are based ultimately on genetic controls. Experimental genetics is the science of dissecting and describing the function of these controls. We can expect genetics to define the components of host-parasite interaction by isolating effects that, as far as we can tell, are due to single genes in host or parasite. A difference between two reactions known to be determined by alleles at a single locus is intrinsically more amenable to explanation than one that has not been so defined. Reason and intuition tell us that it is likely to be simpler than a difference determined by alleles at two or more loci, but even in such terms a complete explanation of the difference between resistance and susceptibility is still a difficult and, in most cases, intractable problem. The biggest obstacle is the unequivocal identification of single-gene effects. However, such an analytical approach has led to our present understanding of the nature of resistance and susceptibility of crop plants. It should, in turn, lead to better ways of controlling the parasites that destroy our crops.

For most of this book I shall be considering higher plants as the hosts. Although their parasites include viruses, mycoplasms, bacteria, fungi, other higher plants, nematodes, and insects, my discussion will be concerned mainly with fungal pathogens. Relatively little work has been done on the genetics of the other plant parasites, but what is known about them suggests that in their interaction with their hosts, all these parasites are remarkably alike. The examination of parasites other than fungi will be to demonstrate this similarity. To describe the interaction we need terms with precise meanings. Some of these are outlined in the following pages. The others I will discuss later in their proper contexts.

SOME DEFINITIONS

The distinction between pathogen and parasite is sometimes useful. A *parasite* lives in or on another organism from which it obtains nutriment. A *pathogen* is a parasite which produces a disease in its host. In this context I define *disease* as a departure from normal metabolism, reducing the normal potentiality for growth and reproduction of the host, caused by the presence of a pathogen. When a host plant exposed to a pathogen is invaded and becomes diseased, we say the host is *susceptible* and the pathogen, or incitant, *pathogenic.* A host plant which does not become invaded and is not diseased is *resistant,* and the potential pathogen is *nonpathogenic* towards that host. We may conveniently distinguish here between two kinds of resistance. One is *nonhost resistance,* such as that shown by wheat to the potato late blight organism *Phytophthora infestans* or by potato to the stem rust organism *Puccinia graminis tritici.* The other, *host resistance,* is the result of genetic modifications of the host which render it resistant to pathogens that would otherwise grow on it. Although nonhost resistance may be more complicated than host resistance, both could in fact be due to the same mechanisms. However, a simple test of this idea will have to wait on our developing the skills needed to cross wheat and potatoes.

This book is concerned almost entirely with host resistance, and rather than use the term nonpathogenic to describe a frustrated pathogen, I will use the term *avirulent.* The term *virulent* will mean ability to produce disease on a resistant host (Day 1960). As I show later, the distinctions between these alternatives are not very clear-cut, and there are degrees of resistance and degrees of pathogenicity. When we examine the differences between resistance and susceptibility, the reason for gradation will become apparent. When resistance is absolute, so that no pathogen development occurs at the expense of the host, it is sometimes defined as *immunity.* In practice, the term is often used if there is no macroscopic symptom of infection following exposure to the pathogen. There is a danger in confusing the term immunity in plants with immunity in animals where it implies interaction of antigens and antibodies. There is little or no evidence of comparable responses in plants (see page 149), and immunity here only refers to a symptomless phenotype. We must always remember that we are describing an interaction between two organisms, and the terms resistant and susceptible describe it only from the host's point of view. To emphasize this point, Loegering (1966) has suggested the terms "low infection type" (to describe a resistant or nonpathogenic interaction) and "high infection type" (to describe a susceptible or pathogenic interaction). The terms "incompatible" and "compatible" also describe these interactions. Such terms are useful since they deal with host and pathogen together. As with the other terms, however, the boundary between them must be defined, and this is not always easy to do.

Parasites belong to two general classes: *obligate* and *facultative.* Obligate parasites cannot multiply in nature without their hosts. For example, viruses cannot multiply outside their host cells. Until recently the rust fungi were considered to be obligate parasites, but some races of certain rusts have been cultured on defined media (see pages 64–66). However, even these are still obligate parasites, since a host plant is necessary for multiplication in nature. No doubt other obligately parasitic fungi, such as the downy mildews (*Plasmopara*), powdery mildews (*Erysiphe*), and other rusts, will be grown in axenic culture as soon as their growth requirements have been established. Facultative parasites, on the other hand, can grow and live on substrates other than living host tissue and can usually be cultured on relatively simple media.

The range of plants which serve as hosts for parasites varies greatly. Many pathogens are highly specialized and limited to a single species or even race of host plant, whereas others are unspecialized and may infect plants in many different genera. Some pathogens, such as certain rusts, complete different phases of their life cycles on unrelated host species which belong to different families or even different phyla. The crop plant or economic host is sometimes known as the *primary host.* The host on which the life cycle is completed is known as the *secondary* or *alternate host.*

METHODS

The detailed methods of studying the genetics of plant parasites and their host interactions are as varied as the organisms themselves. I shall consider them briefly under the three headings of host, pathogen, and interaction.

The Host

The preparation of host material has to satisfy several important criteria: it must be well-grown and vigorous when produced under standard, readily reproduced conditions and available in sufficient quantity. Host plant material is used in two ways: to study variation in resistance of the host itself and to study variation in pathogenicity of the parasite. A plant breeder uses host material in the first way. He must know whether the resistance or susceptibility he observes is a feature of juvenile tissues, such as those found in seedling plants, or of mature tissues, and he must test his plants at the appropriate stage of development. It is also important to know whether the reactions observed truly reflect the host genotype and are not due to abnormal cultural conditions. For example, tomatoes carrying the gene Cf_1 for resistance to *Cladosporium fulvum* are susceptible if grown in short winter days with low light intensities, whereas in long days and sunny conditions they are resistant to appropriate forms of the pathogen (Langford 1937).

The plant pathologist, on the other hand, uses host material in the form of a set of standard varieties to observe the variation among different isolates of a pathogen or the segregation of pathogenicity in a pathogen population. Again, uniformly optimum cultural conditions are important so that comparisons of results obtained at different times and in different places are valid. The host material can consist of germinating seeds, seedlings, and young (or mature) plants or plant parts such as detached roots, leaves, fruits, or even tissue cultures. Culture facilities range from a test tube, the laboratory bench, illuminated incubators, growth chambers, greenhouses, and test plots to the farmer's fields and orchards.

Host material is now maintained in the form of seed stocks and species and variety collections of living plants. In the future some genetic stocks will no doubt be maintained as tissue cultures. Genetic purity is vital and must be safeguarded by controls which eliminate seed admixture, pollen contamination, and other mechanical errors. At the same time, the possibility of mutation must be recognized so that if it is suspected, or likely (some recombinants occur at comparably low frequencies), it can be detected and its frequency estimated by test crosses, adequate replication, and checks (see page 23).

The Parasite

Most of the points mentioned in connection with the preparation and preservation of host material apply equally well to fungal pathogens. There are, however, several special problems. Most obligate fungal parasites must be grown on susceptible host plants. Cross-contamination is a serious problem in maintaining cultures on host plants in a greenhouse because of the difficulty of ensuring complete isolation. For rusts the problems can be minimized by storing uredospores in liquid nitrogen, where viability has been maintained for more than 5 years (Loegering et al. 1966). The spores are collected for preservation by the use of a vacuum-operated cyclone particle collector. Each sample is sealed in a glass ampoule before freezing so that there is no risk of cross-contamination in storage.

Cross-contamination is a minor problem for fungal parasites that can be cultured, but there is the risk that on an artificial culture medium, selection will favor forms adapted to that substrate rather than the host. The consequence is either complete loss of pathogenicity or its reduction after one or more transfers. Fortunately this problem can be largely overcome by preventing mycelial growth in storage. Slant cultures in tubes may be stored at 4°C and protected from desiccation by a mineral oil layer or screw-cap closures. A more convenient alternative which works well for many fungi is to store conidia or other cells, including even hyphal fragments, mixed with granules of dry, heat-sterilized silica gel in small vials (Perkins 1962). The fungi are recovered by transferring a few crystals to a fresh culture medium, and the same storage tube may be repeatedly sampled. Storage in the form of dried infected leaves is useful for such pathogens as bacterial blight of cotton (*Xanthomonas malvacearum*) (Brinkerhoff 1963) or *Helminthosporium* leaf blights of grasses (Nelson et al. 1970). The bacterium may be recovered by placing fragments of dried leaf on agar media, but *Helminthosporium* sporulates on leaf material kept overnight in a moist chamber. Nelson et al. recovered *H. maydis* from dried leaf samples stored for up to 15 years. For the many organisms that withstand it, lyophilization is one of the most convenient and inexpensive methods for long-term storage. It is routinely used by the centers that maintain large collections of genetic stocks of organisms like *Escherichia coli, Neurospora crassa, Aspergillus nidulans,* and *Saccharomyces cerevisiae.* The method has had some use in preserving bacterial plant pathogens.

One of the most effective methods of culture preservation is by freezing at the temperature of liquid nitrogen. The American Type Culture Collection maintains a number of plant pathogenic fungi by this means in addition to the rusts kept as frozen uredospores. Unfortunately, preservation in liquid nitrogen is presently of limited use because it is costly and inconvenient.

Mechanically transmissible viruses may be conveniently stored frozen under liquid nitrogen either in leaf tissue or as purified suspensions. Other viruses must be maintained in their host plants much in the way that obligate parasites were maintained before liquid nitrogen storage provided an alternative.

Nematodes and insects may often be maintained in populations confined, by containers or cages, on host plants grown in the greenhouse. Some can be maintained on culture media. Certain nematodes can be grown on agar substrates with a bacterium as a food source, and others can be maintained on tissue cultures of their hosts.

The Interaction

Disease is induced in the host by inoculation or the introduction of a pathogen. For most fungal pathogens, the conditions which favor maximal rates of infection are a comparatively narrow range of temperatures, relative humidities, and light intensities. Under these conditions each spore, or other propagule, has a given probability of giving rise to an infection. The postinfection environmental conditions determine the rate of growth and development of the pathogen and the host's response to the challenge. When signs and symptoms of disease appear, they are described and recorded. Although it is often sufficient to distinguish only two classes such as high and low infection types, it may be more useful to distinguish more than two degrees of reaction. The leaf reactions of cereals to rusts, for example, are classified into 7 categories. These can in turn be grouped into low (0, 0;, 1), high (2, 3, and 4) and variable (X—the so-called mesothetic reaction) classes to simplify analysis. The scoring of disease reactions in genetic studies is generally more a qualitative than a quantitative evaluation. Pathogen inoculum density can be standardized so that comparisons of discrete lesion phenotypes are possible. As we shall see later, however, the probability of infection, the rate of lesion growth, the final size of the lesions, the degree of sporulation, and a range of other effects on the host are equally important as parameters of the infection.

Inoculation experiments commonly include control plants known to be susceptible to the inoculum. Their reactions confirm that the conditions employed favored disease expression. When the control plants respond in the way expected, the investigator has confidence in the test plant results. It is sometimes necessary to test single plants with several different races of a pathogen to discriminate their resistance phenotypes. When infection is localized, the inocula can be applied to different leaves, different leaflets, or different regions of the same leaf. Alternatively, single plants can be inoculated consecutively, allowing time for symptom expression between successive inoculations. When infection is systemic, the reaction of a single plant to several different isolates may have to be deduced by testing its progeny. An example of this is described later on page 98.

Root and stem infecting pathogens require other techniques. For example, the host plant may need to have its root system or vascular tissue exposed for inoculation. Some wilt diseases can be produced by placing a leafy host shoot in a suspension of pathogen cells. The cells taken up in the transpiration stream grow and produce wilting symptoms.

The use of detached leaves or organs sometimes make possible an increase in the scale of experiments, allowing testing of a wider range of temperatures or light intensities, or a larger number of hosts or pathogens, under conditions where growth-chamber or greenhouse facilities are limited. Some instances are known where the host plant can even be dispensed with altogether, and pathogenic and nonpathogenic isolates may be distinguished by their phenotypes on artificial culture media (see page 75, *Ustilago maydis,* and page 147, *Erwinia aroideae*).

Insect parasites are kept in contact with test populations of host plants by using cages or other barriers in the greenhouse or in the field. Infestation is established by ensuring that environmental conditions are optimum for parasite development. The experience gained in handling greenhouse or laboratory parasite populations is useful here.

GENETICS OF RESISTANCE

Breeding for disease resistance did not begin in earnest until the start of this century. Although the earliest records of differences in reaction to disease between varieties of cultivated plants date back to Theophrastus (372–287 B.C.), one of the first observations of resistance was made by Knight (1799). In England in 1795 and 1796, "blight" (probably caused by *Puccinia striiformis*) was particularly severe, and the only wheats free of the disease were some hybrids that Knight had developed by crossing several different varieties. Darwin (1868) mentioned other examples of varietal differences in reaction among onions to the smudge fungus (*Colletotrichum circinans*), and among grapes and strawberries to the powdery fungi mildew (*Uncinula necator* and *Sphaerotheca humuli*, respectively). However, breeding for resistance did not become the important method of disease control that it is today until two events happened. The first was the general acceptance of the parasitic nature of plant disease, which occurred about the middle of the nineteenth century. An editorial commentary in the *Gardeners Chronicle* of 1852 (page 692) summarizes the situation at that time. The vine mildew, which we know now to be caused by the fungus *Uncinula necator*, had spread to the province of Malaga in Spain:

> The cause of this new pestilence continues as much in the dark as ever; meanwhile opinions are as plentiful as blackberries. Some attribute it solely to the action of the mildew fungus; others assert that the latter supervenes upon a previously diseased state of the tissues. One says the evil arises from badly treating the plant; another, on the contrary, ascribes it to their being too highly cultivated. For ourselves we wait with patience for a better knowledge of the fact, and for the results of experiments.

Experiments with this disease and others soon provided both the answers and the birth of a new science, plant pathology.

The second stimulus was the development of the science, as opposed to the art, of plant breeding, and this took longer to occur. Roberts (1929) has discussed the history of plant breeding prior to Mendel. The foundations were laid by the work of men like Knight, in England, and Joseph Cooper, in the United States. Cooper, a New Jersey farmer, published a remarkable account of practical methods of plant improvement and selection whose significance has only recently been appreciated (Zirkle 1968). J. C. Walker (1951) has traced the steps which led Orton (1905) to select wilt-resistant forms of cotton, cowpea, and watermelon from chance resistant plants in his variety trials. The most important development was Biffen's (1905, 1912) demonstration that resistance to yellow rust (*Puccinia striiformis*) in wheat obeyed Mendel's laws. Working in Cambridge, he showed that the resistance of Rivet wheat was determined by a single recessive gene. Since that time more than a thousand papers have described the inheritance of resistance to the diseases of practically every economically important species of higher plant. Indeed, disease resistance of one kind or another occurs throughout the plant kingdom hand in hand with parasitism. The evolutionary basis for this idea and its implications for agriculture are discussed in Chapter 7.

Breeding for disease resistance in major crops rapidly increased in importance in the early years of this century. The reason for this is clear. Biffen's work had shown that a high level of resistance could be inherited simply. The introduction of genes for resistance was straightforward; protection was "built-in" and did not have to be applied like a fungicide. It was, and still is, a compelling method of safeguarding crops against disease. Resistant varieties appeared to offer the prospect of permanent freedom from disease losses, and perhaps even the elimination of pathogens. But for many diseases this proved to be a false hope (Stakman,

Parker, and Piemeisel 1918). Some resistant varieties failed in a spectacular way, and breeders had to reckon with variation in the pathogen. The search for new and better resistance genes then began and continues today. The knowledge accumulated in this search for resistance and its deployment in crop plants is the subject of this chapter.

TYPES OF RESISTANCE

Resistance to plant diseases and other crop pests can be described either in functional or in genetic terms. The functional terms recognize that resistance may be either highly specific and effective against some parasite biotypes or races, but not others, or nonspecific and equally effective against a range of biotypes. In his book on the epidemiology of plant disease, Van der Plank (1963) introduced the terms *vertical* and *horizontal* resistance to describe the differences in function. The terms were derived from diagrams in which column heights showed the degree of resistance of different host varieties to different races of a pathogen. A variety with only race-specific or vertical resistance had columns that were tall (resistant) and short (susceptible), whereas a variety with nonspecific or horizontal resistance had columns all of the same height whose tops made a horizontal line somewhere between the extremes of susceptibility and resistance. A large number of other terms have been used to describe these two kinds of resistance (Thurston 1971). The two that seem to be most widely accepted at the present time are "general" and "specific" resistance, and they are used in this book. At the present time they can only be used *a posteriori* by observing whether a given example is effective against different forms of a pathogen. There is, as yet, no *a priori* definition and this, as we shall see later, is a serious drawback for plant breeders.

Tolerance is sometimes described as a form of general resistance. It is a property of those varieties which although diseased as severely as others, do not show as much reduction of yield. Relatively little attention has been paid to tolerance as a means of minimizing disease losses, probably because it is more difficult to measure than either specific or general resistance and is far less dramatic. Because they do not interfere with parasite reproduction to any significant extent, the epidemiological consequences of using tolerant varieties are rather different from resistant varieties (Robinson 1969).

The genetic terms that describe resistance describe its mode of inheritance, the subject of this chapter. There are three classes: oligogenic, polygenic, and cytoplasmic.

OLIGOGENIC RESISTANCE

Oligogenic resistance, also known as major gene resistance, is determined by one or a few genes whose individual effects are readily detected. Biffen's example of Rivet wheat notwithstanding, it is much more commonly dominant than recessive. Oligogenic resistance is usually also specific, but this is not always the case. An example of general resistance determined by a single gene is given on page 182. The level of resistance and the time at which it is expressed may vary greatly. Although monogenic resistance is often specific and operative at both the seedling and the mature plant stage, this is not always so. For example, the wheats Thatcher and Marquis possess a specific resistance to race 9 of leaf rust (*Puccinia recondita*), determined by a single recessive gene only expressed by adult plants. Seedlings are susceptible (Bartos, Dyck, and Samborski 1969). In the cereals, adult-plant resistance to rusts is more commonly polygenic and general.

Sources of Oligogenic Resistance

Plant breeders strive to maximize yield and quality. In the course of breeding they eliminate undesirable variation that detracts from this goal, and as a consequence, produce varieties that are genetically uniform. If a new variety is an improvement, it replaces existing varieties and is grown on an increasingly larger scale. Thus, the crop itself tends to become uniform. A consequence of this situation is the development of a pest or pathogen that reduces yield or quality, or both, to such an extent that the variety, or even the crop, is abandoned. The various remedies for this situation include a search for resistance. This usually involves screening a collection of heterogeneous, but genetically related, plants likely to include forms with heritable resistance.

Centers of Diversity

Vavilov (1949) demonstrated the potentially useful genetic variation available in regions of the earth where ancestral or related forms of crop plants occur in abundance either in the wild or as primitive cultivars. These plants have not experienced the intense selection that has produced modern crop varieties. Vavilov originally called such areas centers of origin. It is now known that for a number of crops, the centers of diversity may be far from the areas where wild relatives are found and are likely to have come from materials introduced by migrants and travelers (Zohary 1970). The role of centers of diversity as sources of disease resistance was recently reviewed by Leppik (1970) and Watson (1970b). The value of collecting and maintaining germ plasm from centers of diversity is widely recognized, as shown by the number of national plant introduction services.

For example, the Plant Introduction Service of the U.S. Department of Agriculture (USDA) has regional centers to maintain seed stocks and clonally propagated materials for distribution to breeders and other scientists in public institutions in surrounding states. From time to time collecting expeditions are arranged and plant materials are exchanged with other countries.

By way of illustration let us assume that a plant breeder wishes to locate a source of resistance to a fungal pathogen. Once seed of the host material has been obtained, an efficient screening procedure must be devised. Testing of seedlings or small plants, with only one or two true leaves, allows a considerable saving in space and time. They take up much less room and are ready before mature plants. They are inoculated with spores of the pathogen and placed under conditions which promote infection and disease development. Susceptible control plants ensure that both inoculum and environment favor disease, and that resistant phenotypes are not due to escape from infection or other errors. Under the severe conditions of a greenhouse test, specific resistance almost always shows up best. General resistance is often displayed only by mature plants and cannot be detected in seedling tests. Faced with the choice between high levels of resistance recognizable at the seedling stage and a lower level of resistance, which is sometimes difficult to distinguish from susceptibility even in mature plants, most breeders in the past have selected the first. A survey made by Person and Sidhu (1971) confirms that such choices have dictated the selection of resistance. They classified a total of 912 papers published since 1912 describing disease resistance. Only 60 papers (6.5 percent) described polygenic resistance, and 80 percent of these were published before 1950 when the need for cultures of pathogens of known genetic origin, or standardized for screening, was not widely appreciated. Some of these early reports of complex inheritance could have been the result of using mixed test cultures or different cultures at different times.

Differentiating Resistance Genes

The methods used to incorporate oligogenic resistance are described in textbooks on plant breeding (Allard 1960; Williams 1964) and need not concern us here. However, one technical problem is of special interest, namely, that if resistance to the parasite is already available, it is important to distinguish the new from the old. Two methods may be used. The first makes use of pathogenic races that will eliminate existing resistance genes. In theory, if a race that is pathogenic on all the available varieties with specific resistance is used in the initial screening, only new genes will be selected. In practice, as the number of known resistance genes increases, a series of races are used to test for single resistance genes as they are isolated during the early steps in the breeding program. The underlying assumption is that only resistance that is effective against all known races is both new and useful. Although this has worked as a rule of thumb, it neglects the possibility that genes which are ineffective against some races may still be very useful in areas where these races do not occur (Robinson 1971). The same genes can also be valuable in combination with others. Another snag is that the breeding tests make no comparisons of the epidemiological effectiveness or "strength" of the resistance genes (see page 181).

The second method is a breeding test to see whether the new gene for resistance segregates independently of known genes. This method can be used when a race capable of discriminating a known gene has not been found. For example, it may be needed to distinguish between two more resistance genes found by the first method. The breeding test will not readily discriminate alleles at the same locus, however, and this, as we shall see presently, is not uncommon. Genes for resistance may also be distinguished by differences in their phenotypes when inoculated with the same race. Unfortunately, at the time when identification is most important, before they are used in breeding, the genes to be compared are usually present in quite different genetic backgrounds, which may well have important effects on the phenotype. In some disease resistance programs, the number of identified genes has already become formidable. In wheat, for example, there are more than 20 identified single genes for resistance to stem rust *Puccinia graminis tritici*). In oats a similar number of genes has been identified for resistance to crown rust (*P. coronata*). (Fleischmann and McKenzie 1968). In both these crops, potentially new resistance genes must undergo a considerable amount of testing to prove they do not merely duplicate genes for resistance already in use or discarded in former years.

The literature on the inheritance of resistance to pests and diseases is too large to attempt a survey in this book, but Table 2.1 lists some reviews published in the last 10 years as a guide to further sources of information.

Mutation for Resistance

Resistance may sometimes first be detected under conditions when a disease is so severe that the crop is almost completely destroyed. Orton (1900) selected surviving cotton plants on soil heavily infested with cotton

TABLE 2.1.
Some reviews dealing with resistance to parasites in major
crops during 1960–72.

Crops	Parasites	Reference
Cereals and grains		
Rice	general	Ou & Jennings (1969)
Wheat	powdery mildews	Moseman (1966)
Oats	multiline varieties	Browning & Frey (1969)
Vegetables		
General	insects	Stoner (1970)
Tomatoes	diseases	Walter (1967)
Cucurbits	diseases	Sitterly (1972)
Root Crops		
Potatoes	late blight	Black (1970), Thurston (1971)
Stimulants		
Tobacco	diseases	Burk and Heggestad (1966)
Fruits		
apple	scab	Williams & Kuć (1969)
apple and pear	diseases	Shay et al. (1962)
Forest crops	rusts	Bingham et al. (1971, 1972)
	general	Gerhold et al. (1966)
General	General plant diseases	Day (1968, 1972b), Brock (1967), Van der Plank (1968), Hooker & Saxena (1971), Walker (1965), Kuć (1966), Robinson (1971), Person and Sidhu (1971), Horsfall et al. (1972)
	Insects	Painter (1951, reprinted 1966), Beck (1965), Maxwell et al. (1972)
	Nematodes	Kehr (1966), Bingefors (1971), Rohde (1972)
	Rusts	Hooker (1967, 1969)
	Bacteria	Klement & Goodman (1967), Kelman & Sequeira (1972)

wilt (*Fusarium oxysporum vasinfectum*). Bolley (1905) also obtained flax that was resistant to *F. oxysporum f. lini* under similar conditions. The origin of such resistant plants described in the early literature is not clear, and although spontaneous mutation may be responsible, other sources such as chance outcrossing or seed admixture cannot be ruled out. The criteria used to confirm the mutational origin of resistant seedlings are very exacting (see page 18). In practice, disease resistant mutants of spontaneous origin are usually rare (but see *Sorghum* mutants resistant to *Peri-*

conia circinata, page 19), and such mutants are more profitably sought in material that has been treated with a mutagen. The efficiency of mutant selection depends on rigorously uniform test inoculations so that escapes are not selected as mutants. A common procedure is to treat seeds with either a chemical mutagen or ionizing radiation and then to screen plants raised from the treated seeds (M_1 generation), to select dominant mutants, or to screen their selfed progeny (M_2 generation) to select both dominant and recessive mutants. Chemical mutagens, such as ethylmethanesulfonate, ethyleneimine or diethylsulfate, are convenient for plant breeders without access to a radiation source. Although resistance is often dominant, it is usually profitable to screen M_2 populations to recover homozygous recessives also. In theory, the successful induction of a single mutation in a crop raised from seed requires only the breeding work necessary to produce a true breeding homozygote in an already well-adapted and widely grown variety. In practice, however, as much breeding to remove unwanted variation may be required as in conventional interspecific hybridization (Stephens 1961). Examples of successful induction of disease resistant mutants were listed by Favret (1965) and Sigurbjörnsson and Micke (1969). Murray (1969) recovered mutants in peppermint (*Mentha piperita*) resistant to *Verticillium*. Approximately 100,000 plants, derived from neutron- or X ray-treated stolons, were set out in wilt-infested soil. Some 12 of the mutants had high to moderate resistance and were presumed to be dominant, but this was not tested since *M. piperita* is a sterile allohexaploid ($2n = 72$).

An induced mutation for resistance to *Erysiphe graminis hordei* in barley has been reported to occur with an unusually high frequency. What appears to be the same mutant has been recovered some 9 times in different varieties, in different locations, following treatment with different mutagens. The mutant, called *ml-o* (Favret 1971), is recessive and has the pleiotropic effect of producing a necrotic flecking in the absence of the pathogen. Jørgensen (1971) noted that *ml-o* does not map in the regions of chromosomes 4 and 5 where a number of naturally occurring resistance genes are located. He also noted that plants carrying this resistance were not susceptible to any physiologic race of mildew in large-scale tests in several parts of the world. Unfortunately, the pleiotropic necrotic flecking and an associated small reduction in yield have prevented its commercial use. Presumably the same defects would have eliminated this genotype if it had arisen in wild barleys (but see page 139).

It is of some interest to know whether crop plants with induced mutants for resistance have succumbed to new pathogenic races. Favret (1971) has given 3 examples of mildew resistance in barley (MC 25, Mut 511, and M1-501) that have succumbed. Without evidence to the contrary, it is possible that these mutants are contaminants but it seems unlikely that induced resistant mutants would be much different from naturally occuring ones.

Although, to date, more resistant mutants have been recovered after treatments with X rays and thermal neutrons than with chemical mutagens, this no doubt only reflects the greater use of the former mutagens in large-scale experiments.

VICTORIA BLIGHT—*Helminthosporium victoriae*

One of the most intensively studied systems is the victoria blight disease of oats. The history of this disease and some other features of the causal organism *H. victoriae* are given on page 175. The oat variety Victoria and its derivatives are peculiarly susceptible to *H. victoriae*. When the pathogen is grown in liquid culture, a toxin produced by the mycelium accumulates in the culture medium. The toxin was given the trivial name victorin. It is highly specific even at high dilution and kills germinating seeds of Victoria oats. Victorin has been partly characterized and has a polypeptide fragment of some 5 or 6 amino acids linked to a nitrogen-containing sesquiterpene. Its molecular weight is between 800 and 2,000. Toxin concentrations lethal to plants of the variety Victoria have no effect on plants of oat varieties resistant to *H. victoriae*.

Sensitivity to victorin appears to be associated with a dominant gene (*Pc-2*) for resistance to crown rust (*Puccinia coronata*). Oats homozygous for the recessive allele (*pc-2*) are relatively insensitive to victorin. Attempts to combine the rust resistance of *Pc-2* with resistance to *H. victoriae* were not successful. *Pc-2* is either pleiotropic or else very tightly linked to a dominant gene for susceptibility to *H. victoriae*, called *Hv*. Even the mutational evidence, summarized below, does not enable us to say with certainty whether *Pc-2* and *Hv* are one and the same gene.

The specificity of victorin suggested another method of combining resistance to crown rust and victoria blight. This was to treat large numbers of germinated seeds with victorin to detect spontaneous resistant mutants. Some 502 resistant seedlings were recovered in tests of about 5×10^7 grains of two Victoria derivatives (Wheeler and Luke 1955; Luke et al. 1960). These were checked for crown rust resistance and agronomic type, and 72 were found that were only partially sensitive to toxin, resistant to crown rust, and of the same agronomic type as the two parent varieties. The remainder were either resistant to toxin (but crown rust susceptible) or resistant to both (but of a different agronomic type). Both classes in the remainder were discarded on the assumption that they were either outcrosses or came from mechanical mixtures of other varieties. Progenies of the 72 selected were tested, and only two showed a pattern of reactions to five crown rust races similar to that of the two parental varieties. *Pc-2* gives a characteristic necrotic reaction to certain races. Crown rust resistance in the 70 other selections differed qualitatively and was most likely due to other loci controlling rust resistance, since Victoria has at least two other genes for resistance to crown rust (Simons et al. 1966). These mutants are most easily explained as deletions. Genetic tests of the two spontaneous mutants with partial sensitivity to victorin, and necrotic resistance to rust, showed them to be alleles of *Hv* with the following order of dominance: *Hv > Hvmut > hv* (Luke et al. 1966). In each case dominance was incomplete as the sensitivity to victorin in each F_1 (*Hvmut/hv*) was intermediate. Simons (1970) has reviewed evidence for the existence of other alleles at the *Pc-2* locus among oat cultivars from South America. These are all susceptible to *H. victoriae* but vary in their reactions to different races of *P. coronata*.

The elimination of pollen or seed contaminants by comparing agronomic type or rust resistance is laborious and subject to error. An alternative is to compare the leaf proteins of mutants with those of their putative parents by disc electrophoresis. Wheeler et al. (1971) used this test to show that the isoperoxidases of 3 of the 72 mutants were indistinguishable from those of their putative parents. Since the isoperoxidases of closely related oats (sister selections) were clearly distinguishable in tests run at the same time, they concluded that the 3 were not contaminants.

The victorin screening procedure is one of the most efficient yet discovered for detecting resistant mutants. Konzak (1956) has used it for selecting resistant seedlings from much smaller samples of seed produced by oats exposed to thermal neutrons and X rays. In later studies, Wallace (1965) determined induced mutation rates at the *Hv* locus by treating seeds with gamma rays from a cobalt-60 source and with the chemical mutagens ethyleneimine and diethylsulfate. Mutants were detected as homozygous recessives in M_2 progenies. Seed moisture content had a pronounced effect on the mutation rate produced by gamma radiation.

Luke and Wallace (1969) tested a group of 25 induced mutants with intermediate resistance to victorin for seedling root inhibition at different temperatures. Sixteen showed victorin inhibition approaching the level of the susceptible control at 26°C or below but were highly resistant at 30°C (Figure 2.1). The other mutants showed peak sensitivity at 22° or 26°C, with greatest resistance at 18°C or at 18°C and 30°C. The authors suggest that the different temperature sensitivities may represent mutations in different subunits of the gene. Meiotic products of a cross between mutants in classes E and D in Figure 2.1 could perhaps be screened by crossing the hybrid with an *hv* homozygote and testing the seedlings at 26°C. Non-

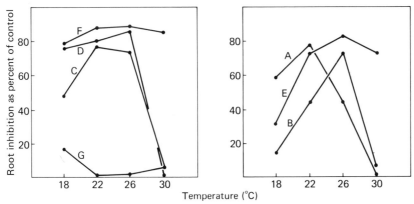

FIGURE 2.1

Influence of temperature and victorin (0.1 unit/ml) on root inhibition (percent of control) of mutants of *Avena byzantina* Koch at the *Hv* locus. The 25 mutants were separated into five classes. A representative of each class is shown: class A, sensitive to the toxin at 22°C; class B, at 26°C; class C, at 22° and 26°C; class D, at 18°, 22°, and 26°C; class E, at 22°, 26°, and 30°C. Curve F represents a susceptible variety, and cuve G represents a resistant mutant. (Luke and Wallace 1969.)

recombinant *HvE* or *HvD* gametes would give susceptible seedlings, whereas the doubly mutant class of recombinant gametes might give resistant seedlings. It would be of interest to know whether the temperature sensitivity of these mutants occurs at the time of interaction with the toxin or when the host tissue is laid down. In other words, is a seedling grown at 30°C sensitive when tested at 26°C and *vice versa*? In view of the evidence for allelism at *Pc-2*, it would not be surprising if some of the mutants with intermediate resistance to victorin also showed a range of reactions to races of *P. coronata*.

MILO DISEASE OF SORGHUM—*Periconia circinata*

This root rot disease is caused by a soil-borne pathogen that produces an injurious toxin causing leaf yellowing and wilting. Breeding for resistance proved to be uncomplicated when it was discovered that resistant selections could be recovered from standard varieties of sorghum without the need for hybridization. Schertz and Tai (1969) established that the spontaneous mutation rate from susceptibility to resistance was of the order of one per 8,000 gametes. They employed a seedling test similar to that used for screening for victoria blight resistance in oats. Germinated seedlings were placed with their roots immersed in a culture filtrate of *P. circinata* for four days. Surviving seedlings were transplanted and self-pollinated and their progeny tested. One in ten proved to be mutant; the rest were escapes. All

14 mutants recovered were heterozygotes showing an intermediate level of resistance compared with homozygotes. F_2 and F_3 progenies showed segregations expected for single locus control of resistance to the toxin, but the number of loci involved among the 14 mutants was not recorded.

How stable such resistant mutants will be toward physiologic specialization remains to be established. In the absence of differentiating races, tests for identity of independently produced mutants are more difficult than those described on pages 13 and 14.

Dominance

Oligogenic resistance is usually determined by single dominant genes. However, there are a number of examples in which resistance is determined by single recessive genes. These include resistance to yellow rust (*P. striiformis*) in Rivet wheat and several other varieties (Biffen 1905; Lupton and Macer 1962), powdery mildew (*Erysiphe polygoni*) in peas, *Pisum sativum* (Harland 1948; Cousin 1965; Heringa et al. 1969), and victoria blight (*Helminthosporium victoriae*) in oats.

Several examples which appear to involve dominance reversal are known. One of these was recently reexamined by Hooker and Saxena (1971). The gene *Rp3* in corn determines resistance to races 901 and 933 of *Puccinia sorghi*. The homozygote *Rp3/Rp3* is resistant to both races, the heterozygote *Rp3/rp3* is resistant to 901 but susceptible to 933, and the homozygote *rp3/rp3* is susceptible to both. One explanation is that *Rp3* consists of two linked loci, one determining dominant resistance and the other recessive resistance. This was tested by looking for recombinants in a progeny of nearly 5,000. None was found even though a linked marker was used to increase selection efficiency. Hooker and Saxena concluded that if there are two genes in *Rp3*, they must be less than 0.02 map units apart. An alternative explanation is that the difference in reaction is a gene dosage effect. Thus, although *Rp3/rp3* is resistant to race 901, only the homozygote *Rp3/Rp3* achieves the threshold level of gene product required for resistance to race 933. Lupton and Macer (1962) described a similar example of resistance to yellow rust (*P. striiformis*) in wheat. In this case two genes showed dominant resistance to races that were considered less aggressive than those to which they showed recessive resistance. The authors suggested that gene dosage was responsible for the effect.

There are few studies of gene dosage effects on disease resistance. In corn, resistance to northern leaf blight (*Helminthosporium turcicum*), controlled by the gene *Ht*, was greater in monoploids and homozygous diploids than in heterozygous diploids. It was greater still in triploids and tetraploids with three and four doses of *Ht* (Dunn and Namm 1970).

Toxopeus (1957) compared potato seedling clones carrying 0, 1, 2, or 3 copies of the gene *R3,* for late blight resistance, for their reaction to race 0 of *P. infestans.* All plants without *R3* were susceptible, but among the remaining plants there were no significant differences in the level of resistance that could have been correlated with different numbers of copies of *R3.*

Favret (1971) recorded that the recessive, induced mutant allele *ml-o,* governing resistance to powdery mildew in barley, is not expressed in triploids or tetraploids in the presence of one dose of *Ml-o,* the dominant wild-type allele.

That dominance may also depend on genetic background is illustrated by Dyck and Samborski's (1968) studies of the inheritance of resistance to leaf rust (*P. recondita*) in wheat. Two alleles of the gene *Lr2* (*Lr2²* from Carina and *Lr2⁴* from Loros) were separately introduced into the variety Prelude by 6 generations of backcrossing. Each behaved as a single dominant gene throughout. Both backcross lines were then crossed with Thatcher and Red Bobs, and F_2 populations tested with race 1. The results are shown in Table 2.2.

Evidently the Thatcher background altered the expression of both alleles, since *Lr2²* showed intermediate dominance in the heterozygote (0; 2⁺ reaction) whereas *Lr2⁴* behaved as a recessive. Both alleles were fully dominant in progenies from Red Bobs inoculated at the same time.

Complementary Genes

Resistance to stem rust (*P. graminis tritici*) in the wheat varieties Gabo, Lee, and Timstein depends on two complementary genes (Knott and Anderson 1956). Examples of complementary gene action for resistance to crown rust (*P. coronata*) in three different oats including Bond are noted in

TABLE 2.2.
Reaction to race 1 of leaf rust of F_2 populations of two Prelude backcross lines crossed to the susceptible wheat varieties Red Bobs and Thatcher. (From Dyck and Samborski 1968.)

| Hybrid | Rust reaction | | Total | Ratio | P |
	Resistant	Susceptible			
Red Bobs × Prelude *Lr2²*	152(0;)*	54(4)	206	3 : 1	.95–.50
Thatcher × Prelude *Lr2²*	49(0;), 78(0;2⁺)	43(4)	170	1 : 2 : 1	.50–.30
Red Bobs × Prelude *Lr2⁴*	138(0;1–)	56(4)	194	3 : 1	.30–.20
Thatcher × Prelude *Lr2⁴*	66(0;1–)	170(4)	236	1 : 3	.30–.20

*(0;) (0;1–) (0;2⁺) represent different levels of resistance.

Simons et al. (1966). Baker (1966) isolated the dominant complementary genes from Bond in separate lines. The F_2 ratios observed when the two lines were intercrossed (9 resistant; 7 susceptible) were the same as those obtained in F_2 segregations from the cross of Bond by a susceptible variety. The two lines were particularly useful in showing that resistance in several other oat varieties was either complementary to, or allelic with, one or other of the two genes of Bond.

Multiple Allelism

IN FLAX—*Linum usitatissimum*

In the course of breeding for resistance to the rust *Melampsora lini* in flax, Flor (1947) noted that genes for resistance from different sources that mapped at, or very close to, the same point on a chromosome could often be distinguished by their reactions with certain rust races. The different genes were assumed to be multiple alleles at one locus. A total of 26 different resistance genes were assigned to 5 independent loci; K(1), L(12), M(6), N(3), and P(4), where the numbers indicate the alleles at each locus (Flor 1971). Although multiple allelism restricts the number of different resistance factors that can be used in a single diploid plant, even the potentially large number of combinations of alleles at up to 5 loci have not been exploited. Rust resistant flax varieties have generally carried only one resistance gene. However, Flor and Comstock (1971) have developed lines with alleles for resistance at up to 3 loci. Their report described the tests needed to select and verify the genotypes of such lines.

Another way of relieving the restrictions on numbers of resistance genes deployed in a single plant was found when it was shown that different alleles at the same locus recombine at low frequencies during meiosis (Flor 1965a). The most efficient method for detecting recombinants is to cross two different resistant homozygotes (for example, *L1L1* × *L2L2*) and then cross the F_1 with a susceptible variety (*L1L2* × *11*). The progeny of this cross are inoculated with a race virulent on *L1L1* but avirulent on *L2L2*. When the seedlings have been scored, they are inoculated with a second race avirulent on *L1L1* but virulent on *L2L2*. The F_1 gametes which transmit no resistance give rise to seedlings susceptible to both races, whereas the reciprocally recombinant gametes give rise to seedlings resistant to both races.

IN CORN—*Zea mays*

A detailed examination of the fine structure of a gene for resistance was carried out by Saxena and Hooker (1968). The *Rp1* locus in corn (*Zea mays*) has 15 known alleles which, except for *rp1*, determine resistance to common rust (*Puccinia sorghi*). *Rp1* is located on the short arm of chromosome 10 between *Rp5* and *Rp6*, which also determine resistance to *P. sorghi*, and it is close to *Rpp9*, which determines resistance to another rust (*P. polysora*). (See (a) in Figure 2.2.) The 14 alleles for resistance, designated by the letters *a* to *n*, are all dominant to *rp1*, and in every heteroallelic combination tested, resistance is always dominant. Thus the heterozygote $Rp1^a/Rp1^b$ is resistant to all races that the two parental homozygotes resist. The allele *Rp1d*, derived from a Peruvian variety, conditioned resistance to all 59 cultures of *P. sorghi* available to the authors.

Seven different heteroallelic crosses were studied. Saxena and Hooker looked for rust-susceptible recombinants in four crosses and for both susceptible and doubly resistant recombinants in the other three crosses. Again recombinants were detected by crossing the double heterozygotes to plants homozygous for *rp1* and inoculating the seedlings that resulted. Susceptible recombinants were detected by inoculating seedlings with a race against which both parental alleles were effective. The doubly resistant recombinant class was detected among the same progeny after a second inoculation with a mixture of two differentially pathogenic races. Further inoculations of detached leaves and progeny tests were used to confirm the screening results. Seedlings with recombinant phenotypes may arise by mutation. For example, the frequency of susceptible mutants from crosses of homoallelic lines with the susceptible tester was 4 in 17,749, or .02 percent. All seven crosses gave frequencies of susceptible recombinants in excess of this figure. The data are summarized in Table 2.3 and Figure 2.2:(b). The frequencies of susceptible plants from crosses *a-d* and *d-k* were

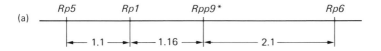

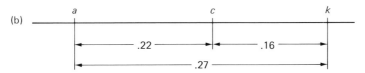

FIGURE 2.2
Linkage maps of short arm of chromosome 10 of corn (*Zea mays*) to show (a) relative positions of loci for rust resistance to *Puccinia sorghi* and (b) positions of 3 alleles of *Rp1* locus. (Based on Saxena and Hooker 1968.)
Note: 2 other allelic pairs, *b-f* and *g-l*, show .10 and .37 percent crossing-over, respectively.
Rpp9 determines resistance to *P. polysora*.

TABLE 2.3.
Summary of heteroallelic recombination data for the $Rp1$
locus in corn. (Saxena and Hooker 1968.)

$Rp1$ alleles	Observed	% recombination
a-k	13 in 9503*	0.27
g-e	11 in 5891*	0.37
a-d	2 in 4170*	0.10
d-k	2 in 5288*	0.08
c-k	32 in 19641	0.16
b-f	14 in 13855	0.10
a-c	32 in 14672	0.22
pooled homozygotes	4† in 17749*	0.02

*Only double recessive (susceptible) phenotype looked for.
†Presumed to be mutants.

considered to be not significantly different from those arising by spontaneous mutation or deletion. Saxena and Hooker suggested that $Rp1^d$
might combine the properties of $Rp1^a$ and $Rp1^k$, thus explaining both its
wide spectrum of resistance and their failure to detect a frequency of
recombinants in a-d and d-k substantially greater than could be explained
by mutation. As noted earlier, doubly resistant recombinants offer plant
breeders an opportunity to increase the number of allelic resistance factors
that they can deploy in a variety. However, in order to detect such recombinants, the breeder must have races that differentiate the parental alleles.
The chief value of the recombinant host gene lies in the fact that an
equivalent recombinant pathotype may not exist. As we shall see, this is
likely to be a very short-lived advantage.

ONE GENE OR SEVERAL?

The question arises whether multiple alleles for disease resistance, such
as the $Rp1$ series, represent mutations in one complex gene or in different
genes which may have arisen either by tandem duplication or by the evolution of close linkage. Since the alleles are separable by differential races,
it is tempting to assume they represent entirely different functions.
However, in the fungi, independently occurring mutations in the same gene
seldom give the same mutant allele. They may (1) show different reversion
rates, (2) produce different kinds of near wild-type revertants, (3) show
different complementation relationships, and (4) show low frequencies of
recombination to produce nonmutant progeny, whereas crosses between
strains carrying the same mutation yield none or several orders of mag-
nitude fewer "recombinants" (Fincham and Day 1971). The fact that

alleles of *Rp1* undergo recombination does not mean that each "allele" represents a separate functional gene. Mayo (personal communication) has pointed out that the 95 percent confidence limits on the largest interval observed in *Rp1* (namely, *g-e* 0.37 percent with limits 0.27–0.53 percent) are such that *Rp1* might well overlap *Rp5* and *Rp6* on either side and also include *Rpp9,* a gene for resistance to another rust-producing species. This region of the short arm of chromosome 10 in maize has undoubtedly evolved so that it is primarily concerned with the control of rust resistance, but until more is known about the nature of gene action in resistance, it seems premature to categorize its organization exactly.

Shepherd and Mayo (1972) recently reported some results of a test to discriminate between one gene or two closely linked genes. The test is based on the expectation that if two closely linked alleles are in different genes, the phenotypes of the *cis* and *trans* double heterozygotes should be the same. The distinction between these two kinds of double heterozygotes is shown in Figure 2.3. If the alleles are in the same gene, the *cis* heterozygote may or may not have the same phenotype as the *trans* heterozygote. If the phenotype is the same the test is equivocal, but if different it suggests that both mutants are present in the same functional gene. The reasoning is as follows: each functional gene directs the synthesis of a polypeptide, and we may assume that the recessive allele for apparent universal susceptibility is due to a null condition, perhaps the formation of

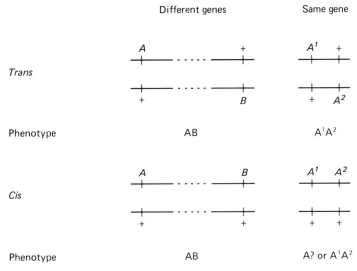

FIGURE 2.3
Comparison of *cis* and *trans* heterozygotes of genes for rust resistance in flax. (After Shepherd and Mayo 1972.)

a polypeptide with no activity. Active alleles are dominant, and they are expressed independently in *trans* heterozygotes. However, in the *cis* heterozygote one of the two segments of DNA responsible for directing polypeptide synthesis now has codons which normally determine two different specificities. If both can be expressed so that the polypeptide performs both functions, the *cis* and *trans* phenotypes will be the same. If neither is expressed, the *cis* phenotype might well be indistinguishable from the null phenotype. With this in mind, Shepherd and Mayo confirmed and extended some earlier findings of Flor (1965a). Reciprocal recombinants were recovered from interallelic crosses at the M locus for rust resistance in flax, but similar crosses between alleles at the L locus only generated doubly recessive recombinants or the null phenotype. Suspecting that plants with this phenotype might be of two kinds, they tested (by selfing) 6 such recombinants from the cross $L^2 \times L^{10}$ to see whether they could recover the original parental alleles. One of the recombinants regenerated the parental allele L^{10}. The nonappearance of L^2 was explained by a supplementary hypothesis. The results suggest that there are structural differences between the L and M series in flax and have important implications for further studies of the nature of disease resistance in other systems discussed in Chapter 5.

POLYGENIC RESISTANCE

Resistance to Yellow Rust in Wheat

Polygenic resistance is determined by many genes of individually small effect. It is usually general, affording resistance to a wide spectrum of pathogen races. The chief features of this form of resistance are well illustrated by Lupton and Johnson's (1970) work on resistance in wheat to the same yellow rust, caused by *Puccinia striiformis,* that Biffen worked with. During 1967–69 Lupton and Johnson grew a collection of old and new wheat varieties representing pedigrees of the leading modern wheats grown in Western Europe. The extent of yellow rust infection on these varieties was scored by estimating the percentage of leaf area infected. Rust epidemics were induced each year in nearby breeding plots by inoculating susceptible plants in "spreader rows" with mixtures of physiologic races that were current at the time. Plants of the pedigree collection became infected following dissemination of the rust spores from the spreader rows. The spectrum of races that were present changed somewhat during the three years. Whereas two were prevalent in all three years, a third appeared in 1969 with a striking effect on some varieties in the collection. The results are shown in Table 2.4. The varieties are arranged in three groups according to when they were introduced. They are further divided according to whether they are susceptible (S), carry specific resistance (SR), or carry general resistance (GR).

The fate of specific resistance among the modern varieties is shown by the susceptibility of Heine *VII,* Nord Desprez, and Rothwell Perdix in all three years. These varieties were resistant in Britain until they encountered virulent races in 1951, 1955, and 1966, respectively. Maris Envoy and Maris Settler were susceptible in 1969 because of the appearance of race

TABLE 2.4
Wheat varieties, arranged according to period of introduction, showing percentage leaf area infected with yellow rust (*Puccinia striiformis*) in 1967–69. (From Lupton and Johnson 1970.)

		1967	1968	1969
Introduced before 1920				
Susceptible	Bon Fernier	75	25	20
	Atle	95	25	30
GR	Browick	0.1	0	0.1
	Little Joss	0–1	1–2	0.1
	Squareheads Master	5	1	0.1
	Yeoman	2–3	1–2	0.1
Introduced between 1920 and 1950				
Susceptible	Desprez 80	95	25	60
	Holdfast	40	5	5
GR	Atle	0	0	0.1
	Bersee	1–2	0	0.1
	Vilmorin 27	2–3	0–1	0.1
Introduced after 1950				
Susceptible	Capelle Desprez	25–30	5–10	5–10
	Heine 51	70	25	30
SR	Heine VII	40	25	20
	Maris Envoy	0	0	50
	Maris Settler	0	0	50
	Nord Desprez	80	25	50
	Rothwell Perdix	90	50	60
GR	Elite le Peuple	0	0	0
	Maris Nimrod	0.1	0	0–1
	Minister	0	0	0–1

GR = general resistance.
SR = specific resistance.

58C. In contrast, the varieties with general resistance supported only low levels of infection in the field in all three years. All of these are susceptible as seedlings, and their resistance is only expressed at the mature plant stage. Varieties with specific resistance are completely resistant as seedlings and mature plants to some races, but are susceptible at both stages to other races.

The inheritance of the general resistance of the old tall-strawed Little Joss was studied by crossing it with the short-strawed Nord Desprez. Although Nord Desprez carries specific resistance, it is highly susceptible to the races used in the tests. Provided that only such races are used in screening segregating progenies, there is no risk of confusing specific and general resistant phenotypes, since only the latter will be expressed.

We may note, in passing, an important corollary. At the time Nord Desprez was selected, recognition of its specific resistance depended on the exclusive use of races unable to overcome it. Any useful general resistance that might have been segregating at the same time went undetected and was therefore lost. When Nord Desprez encountered race 2B in 1951, it was highly susceptible. Van der Plank (1963) has called this loss of general resistance the "Vertifolia effect," after the analogous case with the blight resistant potato variety of that name. It is a common phenomenon.

The F_1 of Little Joss × Nord Desprez grown in 1965, like Nord Desprez itself, was badly rusted. In 1966 an F_2 of 2500 plants was exposed to a severe epiphytotic and 5 percent showed some resistance. Single plant progenies from the resistant plants were tested in 1967, and the distributions are shown in Figure 2.4:a. The few susceptible progenies were most likely the result of selecting susceptible F_2 parents which had escaped infection. An F_4 generation, raised from those F_3 plants with less than 4 percent of their leaf area rusted, was tested in 1968. Resistant single plant selections were again made to produce the F_5 generation tested in 1969, with the results shown in Figure 2.4:b. The levels of infection in F_5 still ranged between those of the parents, but were strongly skewed towards the high level of Little Joss. Whereas some progenies showed a uniform level of resistance, others showed a wide variation in the reactions of component lines, indicating that further segregation for resistance had occurred.

Measurements of straw height in 1967 and 1969 showed no correlation between disease resistance and the tall straw of Little Joss. The authors were able to select F_5 lines, combining high levels of resistance with the short straw and other desirable agronomic characters of Nord Desprez. Lupton and Johnson concluded that the genetic control of resistance in Little Joss was complex. Their tests were made under field conditions with no control of either inoculum density or environmental conditions. Even so, selection worked and promising lines were obtained in F_5.

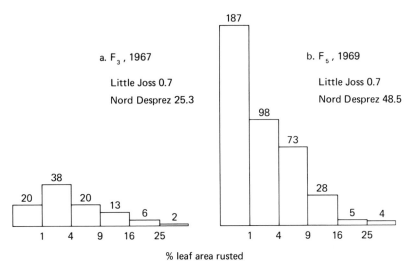

FIGURE 2.4

Number of progenies from single plants of cross Little Joss × Nord Desprez selected as resistant to yellow rust in F₂ and F₄ showing percentages of leaf area infected. (From Lupton and Johnson 1970.)

Resistance to Common Rust in Corn

We have already referred to Hooker's work on race-specific resistance to common rust (*Puccinia sorghi*) in corn. In fact, monogenic race-specific resistance to this rust is only rarely found in commercial corn hybrids in the United States for the good reason that continued selection for general resistance has kept the disease under control. In a study of a large number of inbreds, Hooker (1969) identified only a few that were not resistant. Using them in crosses with other lines with general resistance, he showed that resistance in the F_1 was intermediate, and that transgressive segregation, or the appearance of forms with levels of resistance outside the range of the original parents, occurred in the F_2 and F_3 generations.

Resistance to Late Blight in Potato

No discussion of polygenic disease resistance would be complete without reference to its use in recent years in breeding potatoes resistant to late blight caused by *Phytophthora infestans*. Failure of monogenic race-specific resistance has been documented by Van der Plank (1968) and others. The variability in *P. infestans* that was responsible is described in the next chapter. Breeders were understandably reluctant to turn to polygenic resistance in this case. *Solanum tuberosum* is a tetraploid that is clonally propagated. Russet Burbank (synonym, Netted Gem), one of the most successful North American varieties, was originally released as Burbank's Seedling in 1873. The origin of the russet mutation is unknown. In 1970 this variety made up over 28 percent of the certified seed potato acreage in North America. During the last 40 years many varieties have been released by potato breeders, but most of them account for less than 1 percent of the North American certified seed acreage (Horsfall et al. 1972). In part this is due to the very small chance of recovering better all round phenotypes than those of the few varieties, like Russett Burbank, that dominate the acreage. The conservatism of potato farmers who are naturally reluctant to grow new varieties with cultural and other idiosyncrasies they may not know about is also partly responsible.

New varieties have traditionally been developed by crossing established varieties, or their parent clones, and growing very large progenies which are screened for disease resistance and agronomic characters. The parent lines are highly heterozygous, and very few of the seedlings merit selection and intensive testing. The addition of single dominant genes for blight resistance, largely from the tetraploid species *S. demissum,* was attempted even so, and such varieties as Kennebec and Pungo, which carry gene *R1*, are still grown today, even though physiologic races pathogenic to them are quite common.

Potato varieties with high levels of polygenic general resistance to blight have been successfully developed for markets that are less exacting than those of North America and Europe. These are especially useful in areas where fungicidal protection from late blight is uneconomic. Van der Plank (1963) has observed that varieties probably introduced from Europe in 1833 and maintained in nearly blight-free conditions in the mountains of Basutoland in Africa are extremely susceptible to blight compared with modern so-called susceptible varieties. This suggests that selection has effectively raised the level of resistance since the late blight epidemics of the 1840's.

Several workers have begun new approaches to breeding potatoes. One, at the diploid level, followed the discovery that "haploid" seedlings ($2n = 2x = 24$) of *S. tuberosum* ($2n = 4x = 48$) are formed at high frequency when flowers are pollinated with certain clones of the diploid species *S. phureja* (Hougas and Peloquin 1958). The haploids have been hybridized with wild $2n = 24$ tuberous *Solanum* species, making possible simplified genetic studies with the potato. They have also been crossed with $2n = 48$ cultivars and give some 48-chromosome progeny with agronomic potential (Horsfall et al. 1972).

Another approach, suggested by Simmonds (1960), is to repeat the evolution of cultivated tetraploids from clones of *S. tuberosum* subspecies *an-*

digena, a putative ancestor of present-day cultivars. Starting with seedlings of self- and cross-pollinated origin, selection for tuber size and yield is followed by further selfing and crossing the following season, and so on in alternate seasons. Both approaches make possible a more widespread use of polygenic resistance. In fact, Thurston (1971) has reported that some *andigena* clones have phenotypes for blight resistance which are similar to those oligogenic race-specific resistance genes, and has suggested that this results from the accumulation of a sufficient number of genes of small effect.

How this might work was shown by Estrada and Guzman (1969), who recorded the segregation of resistance in a family of 934 seedlings derived from selfing the moderately resistant *andigena* variety Algodona. The distribution of phenotypes was remarkably close to the theoretical distribution expected if the parent is duplex (*RRrr*) for resistance determined by 4 loci, and shows tetrasomic inheritance where each linkage group is represented four times. Although the 9 most resistant seedlings in the progeny were probably not quite as resistant as the other *andigena* clones noted above, it seems likely that the accumulation of one or two more genes would make them so. Such high levels of general resistance are obtained not just in tetraploids. Graham (1963) noted segregants with unusually high levels of resistance from crosses between resistant and susceptible clones of the diploid Mexican species *Solanum verrucosum* and *S. bulbocastanum* in which the overall pattern of segregation indicated polygenic, not oligogenic, control of resistance. If polygenic resistance of such a high level remains stable and effective, it offers great promise in breeding for resistance. General resistance to late blight in the tomato cultivar Atom was recently shown to be principally controlled by two incompletely dominant genes subject to the effects of modifiers (Grümmer et al. 1969).

The Role of Environment

The importance of environment in the expression and selection of polygenic resistance is illustrated by the experience of Arnold and Brown (1968) with resistance to bacterial blight (*Xanthomonas malvacearum*) in cotton. Working in Uganda they were able to recover useful levels of resistance following several generations of inbreeding and selection in material that had formerly been found uniformly susceptible in Sudan. Unfortunately they were unable to exclude the possibility that differences between the pathogen populations might have accounted for some of the differences in disease development. However, by inoculating duplicate batches of seedlings, they showed that although the unselected population had some resistance in Uganda and was susceptible in Sudan, the selected line was resistant in both countries. The authors concluded that the altered environment enabled them to recognize variation which could be selected and used in that and other environments. Following Waddington (1961) they described this as genetic assimilation.

The techniques of quantitative genetics enable breeders to define the heritability of polygenic resistance and to partition the variation encountered into phenotypic variance, genotypic variance, and genotype-environment interaction. The information can be used to guide a breeder's selection procedures, enabling him to distinguish between heritable and nonheritable variation. The methods were worked out in breeding for characters such as yield, maturity, and quality, which are also in large measure polygenically determined. A full discussion of the techniques of handling quantitative or polygenic inheritance is outside the scope of this book. Interested readers should refer to accounts by Falconer (1960), Allard (1960), and other texts on plant breeding.

TOLERANCE

Common sense would suggest that crop yield and quality will suffer more or less directly in proportion to severity of infection or infestation. Although this is generally true, the behavior of so-called tolerant varieties appears to be an exception. Simons (1966) measured tolerance in oats to crown rust (*Puccinia coronata avenae*) by comparing yield or kernel weight among varieties in split plots. One half plot was rusted, whereas the other was protected by a fungicide. The chief difficulty lay in measuring the relative amounts of infection among the varieties compared. In Table 2.5 some data for five varieties are shown together with rust coefficients obtained by multiplying a numerical score for lesion type by the percentage of leaf area affected. Yield ratios and kernel-weight ratios were obtained by dividing the rusted half-plot values by the protected half-plot values. A ratio value of 1.0 would indicate no effect of rust. The ratios tend to be negatively correlated with rust coefficients, especially in comparisons of resistant and susceptible interactions. Cherokee is significantly more tolerant than Clinton in most comparisons, but then the difference in their rust coefficients leads one to expect this. Evidently the difference between general resistance and tolerance is sometimes difficult to make out. Simons (1969) tested the heritability of tolerance in two crosses, one of which involved Cherokee. In both, heritability was sufficiently high to make selection for tolerance feasible in a breeding program. Brönnimann (1968) has surveyed wheat varieties for tolerance to *Septoria nodorum*. No other form of resistance is known. A recent review by Schafer (1971) discusses some other aspects of tolerance.

TABLE 2.5
Summary of yield and kernel weight data for 5 oat varieties inoculated with 3 crown rust races in 1961. (From Simons 1966.)

Variety	Rust race 203				Rust race 216				Rust race 290			
	Rust				Rust				Rust			
	Reaction	Coeff.	YR	KWR	Reaction	Coeff.	YR	KWR	Reaction	Coeff.	YR	KWR
Clinton	S	37	.663	.754	S	37	.731	.790	S	43	.705	.730
Benton	S	23	.692	.831[a]	S	20	.855	.827	S	12	.754	.852
Cherokee	S	17	.802	.920[ab]	S	10	.732	.885[a]	S	20	.830[a]	.865[a]
Clintland	R	3	.829[a]	.877[a]	R	4	.868	.886[ab]	S	23	.670	.791[a]
Putnam	S	27	.948[ab]	.855[a]	S	40	.745	.831	R	2	.930[ab]	.971[abc]

YR = Yield ratio; KWR = Kernel weight ratio (explained in text); S = susceptible; R = resistant.
[a]Superior to Clinton; [b]superior to Benton; [c]superior to Cherokee at 0.05 level of significance (Duncan Multiple Range Test).

CYTOPLASMIC RESISTANCE

The cytoplasm and its organelles play an important role in heredity, and it would be surprising if there were no examples of cytoplasmically determined disease resistance in higher plants. In fact there are very few. Virus infections sometimes result in immunity to further infection by other forms, and it could be argued that this is a case in point. However, rather few viruses are seed-transmitted, and the efficiency of seed transmission tends to be low (Bennett 1969). Thus not many are likely to show maternal inheritance, the usual criterion for cytoplasmic inheritance. Magaich et al. (1968) described an example where the cytoplasm determines symptom expression in progenies of interspecific crosses of *Capsicum* spp. inoculated with potato virus X. The two species *C. annuum* and *C. pendulum* differ in showing, respectively, a systemic mottle and a necrotic local lesion response after leaf inoculation. The phenotype of the F_1 hybrid depends on which parent is used as female. Here there was no evidence to suggest that virus protection was at work, but it remains a possibility.

The most clearly established case of a cytoplasmically determined reaction to a fungal pathogen is the sensitivity of *Tms* male-sterile corn to race T of *Helminthosporium maydis* (Mercado and Lantican 1961; Villareal and Lantican 1965) (see also page 177). Plants carrying this cytoplasm are susceptible irrespective of whether fertility has been restored. Neither the mechanism of male sterility (see Duvick 1965) nor the nature of susceptibility to the toxin produced by race T is yet understood. Several other sources of cytoplasmic male sterility in corn are not sensitive to race T toxin and have levels of field resistance to southern blight comparable to lines with normal cytoplasm. These other cytoplasms are also designated by letters (C, S, and so on) and are best distinguished from each other by their response to specific fertility-restoring genes. Nelson (unpublished data) has claimed that certain isolates of *H. maydis* are able to severely blight corn with C or S as well as T cytoplasm. However, such isolates have not yet been observed in the field on plants with C or S cytoplasm. Hooker et al. (1970) noted that the general, probably polygenic, resistance of certain corn inbreds to race O of *H. maydis* was also expressed to race T in *Tms* cytoplasm but to a much lower degree. Selection for higher levels of polygenic resistance may well afford an adequate level of protection against race T for plants carrying *Tms* cytoplasm. It seems that until 1969 and 1970 *Tms* cytoplasm in corn offered an entirely adequate level of resistance to other races, but that this was, in effect, specific since it proved inadequate against race T.

It would be of some interest to know whether the cytoplasmic determinant responsible for male sterility is in fact identical with the determinant for sensitivity to *H. maydis* race T toxin. This might be established by examining a range of corn inbreds and varieties to see whether susceptibility ever occurs in consistently male-fertile lines that do not owe their fertility to restorer genes. The material might include lines related to Mexican June from which *Tms* was originally isolated (see Duvick 1965). If susceptibility and male sterility were different, and could be separated, the *Tms* character could still be useful in hybrid seed production.

Another possibility is that *Tms* is the result of either a virus or mycoplasm infection, and that male sterility and susceptibility stem from this. Current work in several laboratories should settle this question before long. Several workers have attempted to cure seeds or seedlings of *Tms* and, more recently, of susceptibility to *H. maydis* by heat and other treatments, but so far, without much success.

The implications of cytoplasmic uniformity have not been lost on plant breeders (see Horsfall et al. 1972). A strenuous effort by many breeders to ensure cytoplasmic as well as genetic diversity among our major crop plants in the future will surely follow. I will return to this topic in Chapter 7.

3

GENETICS OF PATHOGENICITY

During the early years of this century, as breeders began to use resistance to control crop diseases, it became clear that the pathogens could respond in kind (Stakman, Piemeisel, and Levine 1918). The result was sometimes erroneously described as the breakdown of resistance. Resistance did not break down. It was still effective against the original form of the pathogen, but was not effective against new forms. Since the new forms were identical in appearance to the original and differed only in their host range, they were called physiologic races. They are distinguished by their reactions on a group of resistant varieties known as a set of differential hosts. The set usually includes those resistance genes which are in use to control the disease, and so defines the races that are the most important agronomically. A set may also include resistance genes which are no longer in use because they do not control prevalent races. These enable us to trace the development of new races from older races, and they are useful in race classification (see Chapter 4, page 107 et seq.). The resistance of differential host varieties is by definition specific and is almost always oligogenic. Parasites with physiologic races include viruses, bacteria, nematodes, insects, and higher plants (e.g., *Orobanche*), as well as plant pathogenic fungi. Whether physiologic races of mycoplasms exist is not clear, although there seems to be no reason why they should not occur in response to host resistance.

The discovery of physiologic races was followed by a period when plant pathologists explored and catalogued all kinds of variation in pathogenic fungi to try to understand its basis and hence be in a position to control variation in pathogenicity. The rapid development of fungal genetics stemming from Beadle and Tatum's (1945) discovery of induced auxotrophic mutants in *Neurospora* helped not only to expand but also to redirect this work. As a result there is now much evidence on the genetic basis of variability to show that pathogenic fungi are substantially like the genetically better known saprophytes. This chapter is about the genetic control of parasitism and draws principally on examples of pathogenic fungi for illustration. I have included a short introductory section for the reader who is not already familiar with the elements of fungal genetics and how it differs from the genetics of higher plants. The reader who requires a comprehensive treatment should consult Fincham and Day (1971).

GENETIC SYSTEMS IN FUNGI

Cytology

The fungi are considered to be eukaryotes with well-defined nuclei, each bounded by an envelope. Since the nuclei are very small, cytology with the light microscope is difficult. It is also complicated by the fact that the spindle develops and mitosis takes place inside the intact nuclear envelope (Robinow and Marak 1966). The envelope becomes constricted around the two daughter nuclei at telophase. Some of the most satisfactory evidence on the organization of mitosis comes from studies with light and electron microscopy of the same nucleus, a technique pioneered by Girbardt (1965). Observations made by phase contrast microscopy of living nuclei undergoing division can then be compared with fixed, embedded, and sectioned versions of the same object. In this way interpretations derived from light microscopy can be tested by ultrastructural observations. A similar technique was very successfully applied to the plant pathogen *Fusarium oxysporum* by Aist and Williams (1972). These authors showed that the intranuclear spindle extends between an extranuclear spindle-pole body at each pole and that metaphase chromosomes are separately attached to the spindle microtubules by kinetochores (centromeres) which move towards the poles at anaphase. An electron micrograph of a metaphase nucleus in which several of these features can be seen is shown in Figure 3.1.

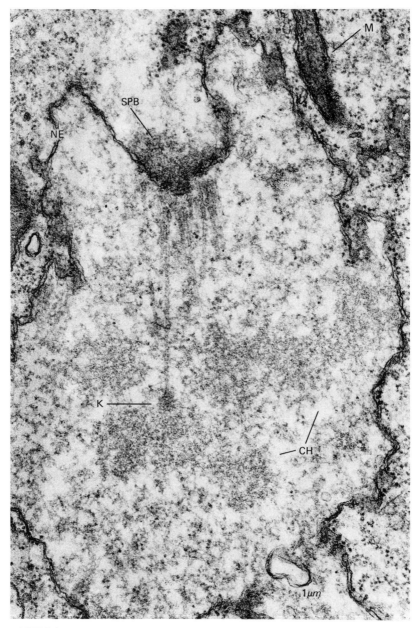

FIGURE 3.1

Metaphase of mitosis in *Fusarium oxysporum* showing a spindle pole body (SPB) sectioned obliquely, and a single microtubule connecting the kinetochore (K) of a chromosome (CH) to the spindle pole. All these structures are inside an intact nuclear envelope (NE) × 82,000. (From Aist and Williams 1972.)

In the absence of comparably detailed comparisons, other interpretations of mitotic organization based only on light or only on electron microscopy, are at best provisional (A. W. Day 1972).

Mitosis in fungi occurs rather rapidly, and in hyphae of *F. oxysporum*, is completed in 5 to 6 minutes (Aist and Williams 1972). In the multinucleate hyphae, a wave of mitosis travels the length of the hyphal tip cell in from 5 to 10 minutes. Although agents such as colchicine can be used to arrest cells of higher organisms at metaphase to facilitate chromosome counts, there are as yet no comparable techniques for fungi.

The peculiarities of somatic mitosis in fungi have only recently begun to be explored. Whether they have genetical implications that need concern us here is still not clear.

In the majority of fungi with an alternation of generations, the diploid phase resulting from fertilization is generally short lived. The diploid nucleus undergoes meiosis without intervening mitotic divisions. Meiotic division figures in the several species of fungi that have been thoroughly studied are much like those of higher organisms. Inheritance is simple since the extended haplophase is comparable to a gamete of a higher plant. Dominance and other genetic interactions can be studied in cells which contain two or more different nuclei (see heterokaryons) or in synthesized diploids that do divide mitotically. In the Oomycetes there is now evidence from DNA measurements that several forms are predominantly diploid, and that meiosis occurs at gamete formation. Many Basidiomycetes possess an extended growth phase, the dikaryon, in which the cells are regularly binucleate. The two nuclei of each cell are derived from the two haploid parental strains which fused to form the dikaryon. The genetics of a dikaryon follow the same pattern as diploid inheritance, even though the diploid stage does not usually occur until just before meiosis.

Sex

In those fungi with a sexual stage, meiosis promotes genetic recombination through random assortment of chromosomes into haploid sets and through crossing-over. Several kinds of controls regulate breeding behavior in fungal populations. The extent of outcrossing can be close to zero in homothallic fungi and approaches 100 percent in the higher Basidiomycetes which have a two-locus, multiple allele system of some complexity for determining mating type. Between these extremes are simple systems with two mating types determined by alternative alleles at a single locus. The frequency of crossing-over between homologous chromosomes is also known to be controlled by loci whose effects are restricted to certain chromosome regions (Simchen and Stamberg 1969).

The extent of our knowledge of sexual mechanisms in plant pathogenic fungi ranges from complete ignorance in the so-called imperfect forms where a sexual stage has not been found, through some uncertainty as to where in the life cycle meiosis occurs, as in *Phytophthora,* to a level where we can say with some assurance that *Venturia,* and several other Ascomycete parasites, are essentially similar to *Neurospora.*

Markers

Most genetic studies require easily scored characters, or markers, whose inheritance can be followed. Pathogens isolated from crops may vary in pathogenicity, mating type, and morphology, all of which can be useful markers. However, naturally occurring markers are rarely sufficient by themselves for genetic analysis with any degree of sophistication. This deficiency is usually remedied by treating spores of a single wild type with a mutagen and looking for various kinds of markers among the survivors which, except for the induced mutation, are isogenic. The most useful classes are auxotrophs that require the addition of a single chemical before they will grow on the simplest defined medium that supports vigorous growth of the wild type (minimal medium), mutants that resist concentrations of drugs that are lethal to the wild type, and pigment mutants. Such mutants can be employed in selection methods where the investigator wishes to focus attention on a very small fraction of a population. These methods require a considerable investment of research effort in basic genetic studies of one or a few genetic stocks. One of the greatest obstacles to detailed studies of the genetics and nature of pathogenicity controls is the lack of concentrated work on one host-parasite system in which the vital shortcuts and special methods, such as are available to *Neurospora* or *Aspergillus* geneticists, can be assembled. The smut *Ustilago maydis* comes near to this ideal because it has attracted some attention as a genetic tool in its own right.

Heterokaryosis

Many fungi can carry two or more genetically different nuclei in a common cytoplasm. This condition is known as heterokaryosis and often occurs following hyphal anastomosis between different mycelia or by mutation within a homokaryotic mycelium. The balance of nuclear types in a heterokaryon may be quite varied. At one extreme are the very stable dikaryons of heterothallic Basidiomycetes made up of regularly binucleate cells. Many Ascomycetes and imperfect fungi have multinucleate cells, and when they are heterokaryotic, the proportions of different nuclei may change in response to selection (Parmeter et al. 1963; Davis 1966). No doubt similar changes occur in the coenocytic hyphae of Phycomycetes. At the other extreme are fungi made up for the most part of uninucleate cells. In these forms heterokaryosis will be transitory and may even be entirely restricted to cells which have come together by anastomosis. When a heterokaryon produces conidia (particularly if they are uninucleate), the segregation of dissimilar nuclei in the spores generates variation. Hyphal anastomosis and heterokaryon formation occur only between individuals which have similar alleles at one or more loci that regulate heterokaryon compatibility (Caten and Jinks 1966; Caten 1971). In the Phycomycetes there is little evidence for the occurrence of anastomosis apart from gamete fusion at sexual reproduction. Demonstration of heterokaryosis in this group involves manipulative techniques that seem unlikely to have a counterpart in nature. Heterokaryosis in the fungi therefore seems likely to be more limited in nature than was at first thought. The restrictions on heterokaryon formation no doubt protect the identity of the organism in much the same way as isolating mechanisms restrict genetic exchange through fertilization. They also may protect mycelia from infection by invasive cytoplasmic determinants or viruses (see page 88). Heterokaryon incompatibility is not necessarily a barrier to sexual outcrossing in *Aspergillus nidulans* (Butcher 1968).

The Parasexual Cycle

Heterokaryosis is sometimes followed by fusion of unlike nuclei leading to the formation of diploids. Diploid formation can be explored in the laboratory by using induced mutants along with selective media, and methods which ensure that heterozygous diploid cells can be recovered efficiently. A number of studies have shown that mitotic recombination and random assortment by haploidization may occur in diploid strains either spontaneously or, at increased frequencies, following treatment with various agents. The phenomena make up what Pontecorvo (1956) called the parasexual cycle. Its incidence among plant pathogenic fungi was recently reviewed by Tinline and MacNeill (1969). Demonstrating the parasexual cycle in nature is difficult because methods for handling markers that may be available have hardly begun to be developed. Thus, diploid strains of *Verticillium* were first tentatively identified on the basis of the size of their conidia. Ingram (1968) treated a large-spored variant of *V. dahliae,* isolated from parsnip root, with the drug parafluorophenylalanine and recovered variant colonies with smaller conidia than the parent strain. This drug is known to induce haploidization in diploid strains of certain other fungi (Lhoas 1961), but it is also mutagenic (Lewis and Tarrant 1971). Complementary auxotrophic mutants were induced in one of the lines with small conidia, and when mixed together, gave rise to prototrophic strains with large conidia. The results strongly suggest that both the large-spored prototrophs and the original large-spored isolate from parsnip root were diploid. Tolmsoff (1972) has also observed that strains of *V. alboatrum* from cotton occur as large- or small-spored forms and claims that these are diploid and haploid, respectively, and may have remarkably different properties.

Among those fungi where sexual reproduction is unknown (Fungi Imperfecti) or of rare occurrence, it has generally been assumed that the parasexual cycle is important in generating and storing variation.

Both sexual and parasexual recombination offer methods of mapping genes and observing their fine structure. Many fungi lend themselves to such studies because of their short generation times, the ease of culturing very large populations, and the availability of efficient selection techniques. With some organisms (such as *Ustilago maydis, Helminthosporium (Cochliobolus) sativum,* and species of *Fusarium*) both systems are available. With others—such as *Venturia inaequalis* or *Phytophthora infestans*—only sexual mechanisms are available, or with still others—such as *Verticillium* or *Ascochyta* (Sanderson and Srb 1965)—only parasexual mechanisms have been used. So far relatively little work on the fine structure of genes concerned with pathogenicity or virulence has been carried out.

Cytoplasm

Just as the different nuclei of a heterokaryon may be partitioned in different conidia to produce variation in vegetative progeny, so may cytoplasmic determinants undergo reassortment to produce a similar effect. Although some elements have been identified, and like yeast mitochondria, may be the vehicle for several linked markers (Wilkie 1970), others are as yet hypothetical and are based on the type of phenotypic segregations observed. Several examples of cytoplasmic determination of growth rate, tolerance of poisons or unusual carbon sources, ability to synthesize lethal substances, and pathogenicity have been demonstrated in fungi. The methods most widely used to demonstrate cytoplasmic inheritance in fungi include maternal inheritance (or, where this is not possible as in the smuts or yeast, irregular or no segregation at meiosis) and the heterokaryon test (Jinks 1966).

Early Genetic Studies of Pathogens

The earliest studies of the inheritance of pathogenicity were made with the rusts and smuts. This now seems surprising in view of the technical difficulties imposed by their obligate parasitism and the complexity of the rust life cycle. Stem rust resistance had become crucially important in wheat breeding by the mid 1920's, but its failure, because of the appearance of physiologic races, made it imperative to gain some understanding of genetic variation in rusts.

Waterhouse (1929) and Newton et al. (1930) observed genetic segregation for pathogenicity in the progenies of selfed races. The first crosses between different races showed that dikaryons were frequently heterozygous, and that pathogenicity on a given resistant variety was very often under the control of a single gene (see Johnson 1953). Genes for pathogenicity were more often recessive than dominant in the dikaryon. One of the first genetic studies of pathogenicity was that of Goldschmidt (1928), who worked with the anther smut fungus (*Ustilago violacea*). He hybridized races which differed in their ability to infect different species of several genera in the family Caryophyllaceae and observed that host-range differences appeared to be under single-gene control. Later studies in the cereal smuts showed that teliospores from nature were frequently heterozygous for genes controlling host range, and that when compatible meiotic products of a single teliospore were intercrossed, the host range of some combinations exceeded that of the parents.

Before proceeding there is one important factor that needs to be stressed. Just as disease resistance in the host plant may be dependent on polygenes, genetic background, and the environment, the same is also true of pathogenicity in the parasites themselves. The breeding of disease-resistant crops has introduced an important arbitrariness into genetic studies of pathogens. Many pathologists have paid far more attention to major gene effects than to the minor gene effects. Although more subtle, the latter are no less important in determining pathogenicity. Pathogenicity, like resistance in the host, is subject to oligogenic, polygenic, and cytoplasmic controls.

The operation of breeding resistant crop varieties in itself generates much knowledge of the genetics of resistance, and at the same time creates a frame of reference for crop parasites. On the other hand, the counterpart in parasites (the production of new races) is a natural phenomenon, and knowledge of the genetics of pathogenicity is harder to come by. Even so, the literature on genetic control of pathogenicity in fungi is extensive, although it is much less so in viruses and insects and is almost nonexistent for plant pathogenic bacteria. It has also been followed by controlled crosses in nematodes. Rather than attempt an exhaustive review, I shall discuss some representative examples drawn from the chief groups of fungal plant pathogens, concluding with a brief discussion of what is known about genetic controls in the other major plant parasites.

PARASITIC FUNGI

The fungi are divided into 4 major taxonomic groups: Phycomycetes, Ascomycetes, Basidiomycetes, and Fungi Imperfecti; and each group includes many pathogens. The summary of genetic systems stressed the fact that fungi have many features in common, but the examples of pathogens discussed in the pages that follow are chosen to illustrate the variety of special problems and features found among them.

I have tried to include enough detail of life cycles and methods of study to enable the nonmycologist to follow the account. For general information he should consult mycology or plant pathology texts.

Phycomycetes: Oomycetes: *Phytophthora*

Phytophthora infestans, cause of the late blight of potatoes, is one of the most important species in this genus, and it has continued to be a hazard to potato growing since it was recognized as the cause of the epidemics which swept eastern North America and Western Europe in the 1840's. Late blight still seriously limits potato growing in parts of South America and Mexico where the climate favors its spread, and control by fungicides is uneconomic. Several other species of *Phytophthora* that have been investigated genetically include *P. capsici, P. drechsleri,* and *P. cactorum.*

The aerial hyphae of many *Phytophthora* species form sporangia that are wind-borne or splash-dispersed. They germinate in water to form about eight free-swimming zoospores that are biflagellate, usually uninucleate, and naked. The zoospores round off forming a cell wall (encyst), and then develop germ tubes which produce appressoria to aid in penetration of the host tissue directly. In the case of leaf pathogens, penetration sometimes occurs through stomata. Growth and development within the host tissue is followed by sporulation and further dispersal. Under favorable conditions, spread through a crop is both rapid and devastating.

SEXUALITY

The species *Phytophthora cactorum, P. megasperma, P. sojae,* and *P. syringae* are homothallic. Genetic analysis of *P. cactorum* has so far been limited to examination of oospores of selfed origin. Shaw and Elliott (1968) recovered several mutants of *P. cactorum* resistant to up to 500 μg/ml streptomycin. Another mutant was streptomycin-dependent. Oospores of selfed origin of both classes of mutant were germinated but showed no segregation (but see page 49). Homothallism introduces the difficulty of distinguishing between oospores of selfed and crossed origin. Although several techniques are available in other fungi, the lack of appropriate markers has so far prevented their application in *P. cactorum*.

The heterothallic species of *Phytophthora*, which include *P. capsici, P. drechsleri,* and *P. infestans,* have two mating types designated A1 and A2 (Savage et al. 1968). Mycelia of both types differentiate antheridia and oogonia (see Figure 3.2). However, isolates of the same mating type vary in their relative sexuality. A given isolate may form oogonia in a mating with one strain, antheridia with another, or both kinds of sex organs with a third (Galindo and Gallegly 1960). Brasier (1971) recently showed that volatile material from cultures of the common soil fungus *Trichoderma vi-*

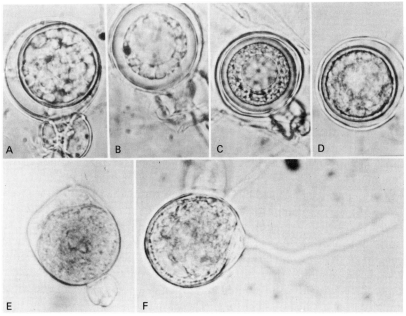

FIGURE 3.2

Oospores of *Phytophthora drechsleri* at different stages leading to germination. A, B—after fertilization (shrunken oospore inside oogonium and empty antheridium below). C—dormant stage showing thickened oospore wall. D—oospore that will germinate in 2 to 5 days showing erosion of inner wall. E—germination. Oospore swollen to fill oogonium and germ tube through oogonial wall. F—elongated germ tube. (From Zentmyer and Erwin 1970.)

ride stimulates the formation of oospores in A2 isolates of seven different heterothallic species of *Phytophthora*. Perhaps the A2 mating type requires a hormonal stimulus from A1 to initiate sex organ formation. The fact that interspecific sexual interactions occur so readily among heterothallic *Phytophthora* spp. could be explained by a common hormonal mechanism. Whether further controls ensure that A2 remains self-sterile, to promote outcrossing, is not known at this time. Sansome (1970) has suggested that some selfing of the A2 parent may well occur in intraspecific crosses of *P. drechsleri,* thus explaining some of the anomalous segregations obtained earlier by Galindo and Zentmyer (1967).

Although *P. infestans* can be grown on a defined culture medium, attempts to recover auxotrophic or other induced mutants have been largely unsuccessful. Clarke and Robertson (1966) and McKee (1969) were not successful in inducing changes in pathogenicity from one physiologic race to another. In *P. drechsleri* and *P. capsici,* Castro et al. (1971) recovered auxotrophic mutants following ultraviolet irradiation of encysted zoospores.

DIPLOIDY

Where meiosis occurs in the life cycle of species of *Phytophthora* is still a vexing question. Sansome (1963) has proposed that meiosis occurs in the antheridia and oogonia of *Pythium debaryanum* and *Phytophthora cactorum,* and that these organisms are diploid throughout the asexual stages of their life cycles. An alternative theory is that meiosis occurs at oospore germination, and that the asexual forms are predominantly haploid. To account for the fact that all single zoospore cultures established from a single germinated oospore are almost always identical, it is proposed that three meiotic products abort leaving only one survivor. Although the cytological evidence for meiosis in gametangia (Sansome 1963) was not entirely convincing, Sansome's theory was strengthened by Bryant and Howard's (1969) demonstration that in the Oomycete *Saprolegnia terrestris* the nuclei of antheridia and oogonia contained, before division, four times as much DNA as the gamete nuclei. Vegetative hyphal nuclei contained twice as much DNA as gamete nuclei and were therefore considered diploid. The information was obtained by microspectrophotometric analysis of single Feulgen-stained nuclei, a technique which has also been applied to other fungi (Davies and Jones 1970).

To date, most genetic analyses of the segregation and recombination of spontaneous or induced markers have given inconclusive results. If the heterothallic species are diploid, then one mating type, possibly A2, is likely to be heterozygous. If a proportion of oospores are of selfed origin, many of the segregations obtained could be difficult to interpret. A recent paper by Shaw and Khaki (1971) described segregation ratios from crosses between the wild type and mutants of *P. drechsleri*. The mutants included one of spontaneous origin resistant to parafluorophenylalanine and two resistant to chloramphenicol that were isolated following treatment with N-methyl-N'-nitro-N-nitrosoguanidine (NG).

The ratios obtained for two of the mutants are shown in Table 3.1 and are those expected if the parents are diploid, the resistant parents being homozygous. This last feature is unexpected and was explained by the authors as being the result of mitotic recombination following mutation. The third mutant behaved in the same way. Alternative explanations for the data are possible but somewhat more complicated than the diploid theory. Although not pursued by Shaw and Khaki, possible explanations include the following: The original resistant parents were all A2 and may have undergone selfing, giving only resistant F_1 progeny. The 1:1 ratios would be expected if the parents were haploid and all oospores were crossed, whereas the 3:1 ratios might result from varying proportions of crossed and selfed oospores. Segregations for mating type among F_1 progenies of these and similar mutants × wild type gave nonmendelian ratios, and some progenies included neuters and ambisexual types (Shaw: personal communication).

Similar difficulties were encountered with *P. capsici* by Timmer et al. (1970). They studied the segregation of methionine and arginine requirements, streptomycin resistance, and mating type. Their conclusion that meiosis occurs at oospore germination was based on the recovery of auxotrophic mutants following ultraviolet irradiation of encysted

TABLE 3.1
Crosses of drug resistant and sensitive isolates of *P. drechsleri* and their progeny. (Shaw and Khaki 1971.)

	Progeny Resistant : Sensitive	Expected* ratio	% oospores germinated
pfa-1 × wt	125 : 0	1 : 0	100
F1 × wt	68 : 57	1 : 1	100
F1 × pfa-1	65 : 0	1 : 0	100
F1 × F1	35 : 11	3 : 1	–
Cl × wt	66 : 0	1 : 0	51
F1 × wt	60 : 48	1 : 1	76
F1 × Cl	59 : 0	1 : 0	47
F1 × F1	51 : 17	3 : 1	58

*If parents are diploid and resistant parents are homozygous.
pfa = para-fluorophenylalanine resistance.
Cl = chloramphenicol resistance.

zoospores, and the recovery of four classes of progeny from crosses of the type *arg met*$^+$ × *arg*$^+$ *met*. Unusual segregation ratios were explained by suppressors with effects both on the auxotrophic markers and on streptomycin resistance. The inheritance of mating type was not simple and was explained by a rather elaborate hypothesis involving three independent pairs of alleles. However, oospore germination in the crosses ranged only from 0.1 to 5.0 percent, so that some abnormalities could be attributed to differential viability of the products.

Low germinability of oospores has also hampered laboratory studies of *P. infestans,* but several workers have shown that mating type and physiologic race type recombine in crosses between different races. Romero and Erwin (1969) claimed that the products of a single germinated oospore were the same, but Laviola (1968, in Gallegly 1970) found that some oospores gave rise to two or more different genotypes. A total of 58 oospores from three different crosses gave rise to germ sporangia that produced zoospores. Laviola isolated two or more different products from only 6 of these. Another 28 oospores from the same three crosses gave rise to germ tubes, but no segregation was noted in tests of the cultures derived from these oospores. The *P. infestans* results lend support to gametangial meiosis, but Laviola's findings show that recombination may also occur at oospore germination. However, Sansome and Brazier (1973) recently carried out a cytological examination of gametangia in some of Laviola's strains and found that maturing oogonia occasionally contained two egg cells (oospheres) instead of the usual one. These and other abnormalities, including restitution nuclei and multivalent associations leading to chromosome duplications and deficiencies, could account for Laviola's recovery of more than one genotype among progenies of single oospores.

Homothallic species of *Phytophthora* form zoospores and oospores, so that populations of asexual and sexual origin derived from the same clone can be compared. Variation among the asexual progeny will be caused by mutation, by the parent clone being a heterokaryon, by cytoplasmic variation, and if the parent clone is diploid, by mitotic recombination. The same sources will also contribute variation to the sexual progeny. However, if the parent clone is diploid and heterozygous for some loci, and if oospores result from fusion of different meiotic products, the variation among sexual progeny will be greater than that among asexual progeny. Boccas (1972) found this to be so for *P. syringae.* He compared colony diameters after 5 days growth on potato-glucose agar at 26°C and noted highly significant differences in variation between the two kinds of populations over three generations. If *P. syringae* were predominantly haploid, then variation among oospores derived from fusion of haploid nuclei, followed by meiosis at germination, should not differ significantly from that among zoospores. The greater variation in growth rate among colonies derived from oospores is most simply explained by genetic recombination and reassortment following meiosis at gamete formation.

Most recently Elliott and MacIntyre (1973) showed that zoospores of the homothallic *P. cactorum* treated with NG produced prototrophic mycelia that carried auxotrophic mutants that segregated in subsequent generations. The patterns of inheritance obtained were consistent with a life cycle having a diploid vegetative mycelium in which meiosis occurred in the gametangia.

Progress in this rather difficult area will be more rapid when researchers in different laboratories agree to concentrate on certain stocks of one species and thoroughly explore their behavior by confirming each other's findings. It may be that unusual, or even novel, genetic systems will be uncovered among these fungi, and that not all Oomycetes are alike in this respect. At present we are attempting to force too few results to fit preconceived ideas derived largely from Ascomycetes and Basidiomycetes. It may be that in some Oomycetes, meiosis occurs in the gametangia, whereas in others it occurs at oospore germination.

VARIATION IN *P. INFESTANS*

During recent years, potato breeders trying to incorporate resistance to late blight have switched from monogenic specific resistance to polygenic nonspecific resistance. The reason for the shift was the rapidity with which new races of *P. infestans* appeared on varieties carrying monogenic resistance. The genes were denoted *R1, R2, R3,* and so on, the numbers indicating different specificities. Most of them were derived from the hexaploid wild potato (*Solanum demissum*). These genes are not known to be allelic, although they are numbered in a way that suggests an allelic series. By 1969 eleven *R* genes had been recognized (Malcolmson 1969), but no gene governed resistance to all known races. The races of *P. infestans* are numbered to indicate their virulence towards potato varieties carrying the *R* genes. Thus, race 1,2 is virulent not only on varieties with no resistance genes, but on those with *R1* or *R2* or both genes, and so on (see also page 107).

How the new races arose so readily became a central problem. In Mexico, where both mating types occur, sexual reproduction appears to generate recombinants maintaining a variety of phenotypes virulent even on wild populations of *S. demissum,* the source of many of the resistance genes. The plants survive the wet periods (when blight is prevalent) as underground stolons.

Only one mating type (A1) of *P. infestans* occurs in the potato growing regions of North America and Europe, and yet the fungus appears to be no less variable there. Parasexual recombination mechanisms were sought and some evidence for them was obtained. One method used by several investigators is to mix inocula of two races of the same mating type and apply the mixture to susceptible potato leaves. The sporangia formed after incubation are harvested and either inoculated to fresh susceptible leaves or screened for recombinants. These are recognized by their virulence patterns on tester cultivars resistant to both parental races.

For example, Malcolmson (1970) mixed sporangia of race 3,4,10 with sporangia of race 1,2,3,7, and after several cycles of inoculation, recovered race 1,2,3,4,7,10. The new race was virulent on the clones $R1R2R3R4$, and $R7$ and $R10$, but avirulent on clones carrying $R5$, $R6$, $R8$, and $R9$. Malcolmson reported several other examples in which she was unable to recover recombinants. Denward (1970) carried out similar experiments which are more difficult to interpret. Starting with races 1, 3, and 4, he first established that they were unstable when separately grown on their respective hosts $R1$, $R3$, and $R4$ for 12 successive passages. This is illustrated in Table 3.2. According to Denward the change from race 1 to virulence on $R1$ and $R4$ occurred after the fourth passage, but it was not until the ninth passage that race 1,4 virulent on $R1R4$ was recovered. Similarly, the change from race 3 to race 3,4 was also preceded by the ap-

TABLE 3.2
Sporulation by serial subcultures of *Phytophthora infestans* race 1 maintained on detached leaves of R1, and race 3 maintained on detached leaves of R3. At each passage inocula from R1 and R3 were tested on the host genotypes shown. (From Denward 1970.)

Start:	Race 1				Race 3			
Passage No.	r	R1	R4	R1R4	r	R3	R4	R3R4
1	+	+			+	+		
2	+	+			+	+		
3	+	+			+	+	+	
4	+	+	+		+	+	+	
5	+	+	+		+	+	+	
6	+	+	+		+	+	+	
7	+	+	+		+	+	+	
8	+	+	+		+	+	+	+
9	+	+	+	+	+	+	+	+
10	+	+	+	+	+	+	+	+
11	+	+	+	+	+	+	+	+
12	+	+	+	+	+	+	+	+
Finish:	Race 1,4				Race 3,4			

death. Some have a known perfect stage in the Pyrenomycetes, the best-known genera being *Cochliobolus* and *Pyrenophora*. In nature, perithecia occur on senescent or dead host tissue and may be important as a means of overwintering. Laboratory studies have shown that all species with a perfect stage are heterothallic and bipolar. Many species with no known perfect stage will take part in forming hybrid perithecia but rarely form viable ascospores. Matings are carried out by inoculating compatible strains on either side of a piece of autoclaved host tissue, such as a leaf fragment or seed, on the surface of nutrient agar in a Petri dish. Perithecia form in 6 or 7 days and contain asci with mature ascospores within 22 days.

The ascospores of *Cochliobolus* are long (100–350μ) and narrow (10–30μ) each with 5 to 10 cells. They are twisted around each other within the ascus, making tetrad analysis difficult. Random ascospores are used for genetic analysis and are obtained either by collecting discharged spores or by crushing mature perithecia. The ascospores of *Pyrenophora* and *Trichometasphaeria* are much shorter, but like *Cochliobolus*, are not linearly arranged at ascus maturity.

Tinline (1962) demonstrated parasexuality in *H. sativum* by using strains carrying induced markers to synthesize heterokaryons and diploids. Mitotic recombinants were selected from the diploids. The perfect stage, *Cochliobolus sativus*, is only known from laboratory studies, so that the parasexual cycle could be an important source of variation in nature in this species. Although parasexuality no doubt occurs in the other species, it has not been demonstrated. For the most part these have been subject to a different form of analysis.

HOST RANGE

Since 1956 R. R. Nelson and his colleagues have examined the genetic relationships among a number of the forms of *Helminthosporium* that are parasitic on members of the *Gramineae*. In the course of this work, the perfect stages of several species were described for the first time and the genetic controls of sexuality, mating type, and the fertility barriers between mating groups were explored. These studies showed that many *Helminthosporium* isolates from wild native grasses from different parts of the world are indistinguishable from those causing diseases of economic crops. Although greenhouse tests do not necessarily indicate the field capabilities of a pathogen, it is clear that the forms harbored by wild grasses are potentially dangerous to crops. The wide host range of most *Helminthosporium* species suggests that in spite of the sterility barriers which restrict intercrossing, the population can be regarded as a continuum of forms adapted to many different members of the *Gramineae*. For example, pearance of virulence on *R4* as well as *R3*. Gallegly and (1959) had earlier shown, and others have since confirmε *festans* mutates readily to virulence on *R4*. The surprising aι feature of Denward's work is that changes from race 1 to rε race 3 to race 3,4 were not apparent as soon as virulence oι A simple explanation is that since the potato clones *R1, R4* not isogenic, an additional step was required to overcome resistance factor (*Rx*) in the *R1R4* clone. In other words, ν selected as race 1,4 was in fact race 1,4,*x*.

The plasticity of *P. infestans* towards *R4* suggests that result could be explained by race 1,2,3,7 becoming 1,2 pathogenicity on *R10* was the result of contaminating sporε 4,10. However, against this there is the reported stability binant and the expectation that race 3,4,10 would be elimiι *R2*.

In further experiments Denward (1970) claimed to havε types. For example, mixtures of races 1 + 3 and of races rise to race 1,2,3,4, stable in further cultures from single ε the new specificities appeared as a result of mutation, bination, or some other interaction, perhaps involving tl not known.

The shift to field resistance for control of late blight hι portant successes, and varieties developed in Mexico suc (Niederhauser 1968) have considerable promise for are cultivation depends solely on resistance. The large-scale p in North America and Europe relies on control by sp gicides. Blight forecasting (Bourke 1970), based on w made spraying both maximally effective and economic tection has allowed a variety such as Russet Burbank, resistance to late blight, to predominate in the United S century. It currently makes up some 23 percent of t acreage. Fortunately, fungicide-resistant late blight straι peared to threaten control, but in the meantime industι have come to depend on a few clones which are extremely cate as blight-resistant varieties. These topics are also d ter 7.

Ascomycetes: *Helminthosporium*

The genus *Helminthosporium* includes the conidial stε causing a number of important diseases of cereals, grasε Several species produce extracellular toxins which briι

Nelson and Kline (1969) made crosses among four isolates of *Helmintho-sporium* from corn (*Zea*) and one from canary grass (*Phalaris*) which had in common the production of curved conidia greater than 100μ long. The parents and ascospore progenies were screened for pathogenicity on) un-related grasses belonging to as many genera. When one parent was patho-genic, and the other nonpathogenic, to a given host, the segregation ratios among the progeny in each case suggested that pathogenicity on the host in question was controlled by one or two genes. The properties of patho-genicity on the various hosts appeared to be inherited independently.

These results raise the question whether the differences between grass genera, defined in terms of the host range of *Helminthosporium* isolates, are analogous to the kinds of differences between cultivar genotypes within a host species which define physiologic races. The so-called *formae speciales* of the rust *Puccinia graminis,* or of the powdery mildew *Erysiphe graminis,* which are adapted to different cereals are another example (see pages 69 and 57).

Some forms of *Helminthosporium* are particularly well adapted to host varieties grown on a large scale. During the last 30 years, important epiphytotics on rice, oats, and corn have been caused by species of *Helmin-thosporium.* The occurrences on oats and corn illustrate some important principles and are discussed in Chapter 7. We will examine another aspect of the genetic basis of their adaptation at this point.

TOXIN PRODUCTION

Helminthosporium victoriae produces a toxin (HV-toxin) active, even at high dilution, against oat varieties such as Victoria or Park, which carry the gene *Pc-2* for crown rust resistance (see pages 17–19). Pathogenicity and toxin production are closely correlated. Race 1 of *H. carbonum,* a corn pathogen, also forms a toxin (HC-toxin) which is somewhat less active than HV-toxin but nevertheless host specific. At a concentration of 0.5 μg/ml, HC-toxin inhibits seedling root growth of susceptible corn hy-brids. At 25 μg/ml, it affects seedling roots of resistant hybrids and other species such as tomatoes, oats, and cucumbers which are not hosts. Concentrations of HV-toxin as low as 0.009 μg/ml completely inhibit root growth of susceptible oats, but at 3,600 μg/ml cause only partial in-hibition of resistant oats, corn, and other nonhosts (Kuo et al. 1970). Both HV- and HC-toxins are polypeptide derivatives. HV-toxin alters the per-meability of cell membranes causing rapid leakage of electrolytes. Some other properties of these toxins were reviewed by Owens (1969) and Luke and Gracen (1972).

When *H. victoriae* was crossed with *H. carbonum* and the progeny tested for toxin production, the results were as shown in Table 3.3. The

TABLE 3.3

Toxin production among progenies of three crosses of *Helminthosporium* (*Cochliobolus*) *carbonum* (race 1) X *H.* (*C.*) *victoriae*. (Data from Scheffer et al. 1967.)

Cross*	Number of isolates forming			
	HV toxin	HC toxin	Both toxins	Neither toxin
1 X 2	20	23	16	22
1 X 3	5	4	2	3
1 X 3'	0	23	0	29

*1 = *H. carbonum*; 2, 3 = different isolates of *H. victoriae*; 3' = a spontaneous mutant of 3 which had lost the ability to form HV-toxin.

amount of toxin a given culture formed varied, but whether each toxin was produced appeared to be determined by independent single genes. Tests of pathogenicity showed that only those cultures that formed HV-toxin were pathogenic on Park oats, and only those that formed HC-toxin were pathogenic on susceptible corn hybrids. Cultures that formed both toxins were pathogenic on either host and consequently had a wider host range than the parents. The last line of the table shows the progeny resulting from a cross between *H. carbonum* and an isolate of *H. victoriae* which had spontaneously lost the ability to produce HV-toxin. As expected none of the 52 cultures tested formed HV-toxin. Other studies (Scheffer et al. 1967) have shown that the difference between race 1 of *H. carbonum* and other isolates which form no HC-toxin, and thus do not differentiate between susceptible and resistant corn hybrids, is also determined by a single gene.

The epiphytotic of southern leaf blight on corn in 1970 was due to the appearance, on a massive scale, of a new race of *H. maydis* called race T. Corn plants carrying Texas male-sterile (*Tms*) cytoplasm (see page 177) are peculiarly susceptible to a toxin formed by race T. The form of *H. maydis* common in the United States corn belt prior to 1970 is called race O. It produces a nonspecific toxin and does not differentially infect corn plants with normal and *Tms* cytoplasm. Lim and Hooker (1971) crossed race O and race T and found that among the progeny, toxin production (assayed by measuring root inhibition of *Tms* seedlings) and high pathogenicity to *Tms* plants were correlated and under the control of a single gene. *H. maydis* race T toxin is like HC-toxin in having a much lower activity than HV-toxin.

Several other species of *Helminthosporium* form toxins. Some recent work on one of these, *H. oryzae,* suggests that toxin production may depend on extranuclear determinants. Lindberg (1971) reported a "disease" of *H. oryzae* which he claims is transmitted from infected to healthy mycelia through hyphal anastomosis. Although infected strains grow less vigorously than healthy strains, they cause proportionately more damage to infected rice plants because they produce a toxin. Cytoplasmic determinants with invasive properties are well known in several fungi (Fincham and Day 1971). It is possible that some of these examples will eventually be associated with the presence of viruslike particles. A similar example to *H. oryzae* toxin is shown by current work on the killer factor in *Ustilago maydis,* described on pages 79–80. Mycoviruses have recently been found to be surprisingly common and are discussed in more detail on pages 86–89.

Ascomycetes: *Erysiphe graminis*

The powdery mildew fungi are grouped into six genera in the Erysiphaceae. Their hosts include a wide range of higher plants on which they are obligate parasites. The powdery mildews of barley and wheat (*Erysiphe graminis hordei* and *E. graminis tritici,* respectively) are the best known genetically within the group. The related form *E. graminis avenae* is an important parasite of oats. Kimber and Wolfe (1966) have claimed a chromosome number of 2 in all three forms and a consistent difference in size between the two chromosomes. The conidia of *E. graminis* germinate at very low relative humidities (Figure 3.3:A). On the host epidermis, the germ tube produces an appressorium with an infection peg that penetrates the cuticle and epidermal cell wall (Figure 3.3:C, D). In a susceptible host the infection peg forms a haustorium inside the penetrated cell. The infection spreads through formation of secondary hyphae from the original germ tube. These repeat the penetration sequence and form haustoria in adjacent epidermal cells (Figure 3.3:B). Bracker (1968) has given a full account of the ultrastructure of the haustorium of *E. graminis* in barley. Only the host epidermal tissue is infected, but in spite of this localization, photosynthetic efficiency is drastically reduced. It may, for example, cause losses of up to 39 percent of grain yield in spring oats (Lawes and Hayes 1965). As the process is repeated the mycelial mat spreads and after 7 to 10 days begins to form chains of uninucleate conidia (Figure 3.3:E, F). The powdery appearance of the sporulating lesions gives the group its name. *E. graminis* is heterothallic, and spherical fruit bodies (cleistothecia) about 200 μm in diameter, containing 8 spored asci, are produced in compatible matings on the host. No barriers to hybridization between

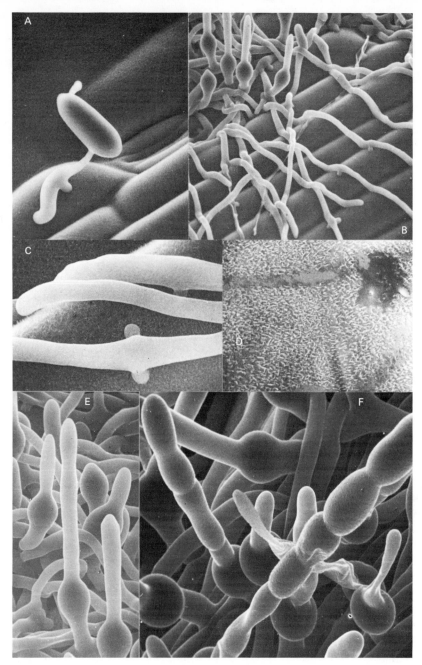

FIGURE 3.3

Development of *Erysiphe graminis hordei* on leaves of susceptible barley. Fresh material as seen in scanning electron microscope. A—germinated conidium at 24 hours showing appressorium and elongating hypha × 600. B—margin of the mycelium at 4 days showing penetration pegs and conidiophore initials × 360. C—ramifying hyphae at 7 days with paired a pressoria × 1,500. D—host cuticle exposed by removing superficial mycelium showing hyphal track in wax crystals and position of appressorium and hole made by penetration peg × 1,500. E—early stages of conidiophore development × 720. F—immature conidia at early (left) and late (right) 4-celled stage. Note localized damage caused by electron beam × 1,010. (From Day and Scott 1973.)

the form species from wheat, barley, rye, or *Agropyron* were found by Hiura (1965). However, hybrid ascospores inoculated to each of the parental hosts were either nonpathogenic or reduced in pathogenicity. In Israel several grasses support infection by both forms, suggesting that hybridization may be frequent in this region, which is one of the centers of origin for wheat, oats, and barley (Eshed and Wahl 1970).

Breeding for resistance to *E. graminis* in wheat, barley, and oats led to the discovery of physiologic races. Crosses between different races are made by dusting seedling leaves with conidia of compatible isolates. These are maintained on appropriate hosts kept in isolation in a greenhouse. Dried mature cleistothecia remain viable for up to 13 years (Moseman and Powers 1957). When moistened the contents swell, cracking open the wall. Ascospores discharged on 2 percent water agar may be transferred singly to seedling leaves, but only about 10 percent give rise to infections. The studies show that virulence on resistant wheat and barley varieties is determined by single genes (Moseman 1966). In a recent review on the genetics of barley mildew, Wolfe (1972) noted that at least 12 loci and 17 alleles determine pathogenicity on resistant barleys. So far genetic studies in *E. graminis* have not included any markers other than virulence or mating type. No methods are yet known for synthesizing diploids or exploring parasexual mechanisms. I am not aware of any successful efforts to obtain axenic cultures of powdery mildews.

Ascomycetes: *Venturia inaequalis*

The apple scab fungus *Venturia inaequalis* invades apple leaf and fruit tissue by penetrating the cuticle and growing just beneath it above the epidermis (Figure 3.4:A). The underlying tissues are affected. If fruits are infected during development, they become cracked or misshapen and unsalable. Heavy leaf infections can also lead to defoliation. In cold climates the fungus overwinters as dormant mycelia in the leaf litter on orchard floors, producing perithecia in the spring which discharge ascospores. Carried by air currents they land on young leaves and produce infections. The disease spreads subsequently by conidia formed by the earlier infections. The ascospores germinate to form haploid mycelia made up of uninucleate cells. Since the ascospores are meiotic products, they vary in genotype, the extent of variation depending on the degree of inbreeding in the population. There are two mating types, and 50 percent of random matings are fertile. Inbreeding is unlikely, since both the ascospore discharge mechanism and subsequent secondary spread by conidia ensure efficient mixing prior to sexual fusion.

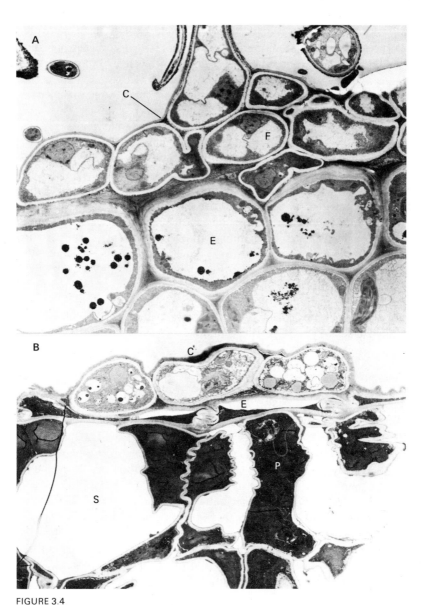

FIGURE 3.4

Ultra-thin sections of apple leaves infected with *Venturia inaequalis* × 3,500. **A:** sporulating colony on susceptible tissue 10 days after inoculation. Several layers of fungal cells (F) lie beneath the cuticle (C) and above the turgid epidermal cells (E) of the host. **B:** Resistant, or fleck, reaction showing necrotic, collapsed epidermal (E) and palisade (P) cells with large intercellular spaces (S). (Unpublished electron micrographs of K. Maeda Nishimoto and C. E. Bracker.)

PATHOGENICITY

Laboratory cultures from single ascospores vary in their ability to produce lesions on the leaves of different apple varieties. Some combinations produce only a pin point or tiny fleck on the leaf surface with no sporulation, whereas others produce large sporulating lesions (Figure 3.5). Two cultures of different mating type, one producing a fleck, the other a lesion on a given variety, can be crossed on an artificial medium in the laboratory, and the eight spores from an ascus isolated (Figure 3.6), cultured, and tested on the apple variety.

For example, two cultures with the host reactions shown in the first two lines of Table 3.4 were crossed. Among 35 asci three types were found,

FIGURE 3.5

Second division segregation for virulence in *Venturia inaequalis* on leaves of the apple cultivar "Haralson." Ascospore pairs 1 and 2, and 7 and 8 are virulent, whereas 3 and 4, and 5 and 6 are avirulent. (Keitt and Langford 1941.)

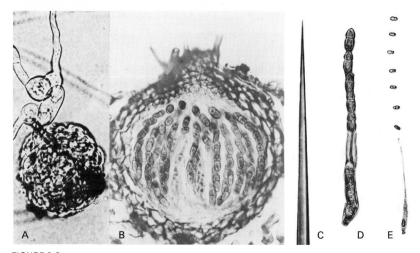

FIGURE 3.6

Sexual analysis in *Venturia inaequalis.* A—protoperithecium showing trichogyne and three antheridia (× 510). B—mature perithecium in section (× 320). C—tip of hand-held glass needle used for spore isolation (× 400). D—ascus ready for dissection (× 400). E—ascus with 7 spores removed and arranged in serial order (× 145). (From Keitt and Langford 1941, and Keitt, 1952. Univ. Chicago Press.)

suggesting that two independent genes controlled virulence on the four varieties. As a result of similar tests, Boone and Keitt (1957) described 7 different genes (*p1–p7*) determining virulence on commercial apple varieties. Bagga and Boone (1968a), using two *Venturia* isolates causing lesions on most of these commercial varieties, but flecks on a set of 9 different crabapple cultivars, identified another 6 genes (*p14–p19*). Williams and Shay (1957) had earlier described 6 genes (*p8–p13*) for virulence on four other crabapple cultivars. The effects of 4 of these genes were somewhat less pronounced than those shown in Figure 3.5. The significance of these studies for interpreting host resistance are discussed on page 105 (see Table 4.10). In North America, infection begins anew each spring from ascospores, and there is little or no carry-over of fungal clones by mycelia or conidia from year to year. Consequently, the genetic pool is so well reshuffled each spring that it is not worthwhile recognizing physiologic races; there would be too many.

FORMAL GENETICS

The genetics of *V. inaequalis* was recently reviewed by Boone (1971). The work of the Madison, Wisconsin, group began under the leadership of G. W. Keitt, and over a period of some 30 years, has explored the genetics of this organism in great detail. Some 30 or more markers have been mapped, including many of the 19 pathogenicity loci. Attempts to force heterokaryons with auxotrophic and other markers have not been successful, and so there is no information on parasexuality. In view of the uninucleate nature of individual mycelial cells, it may seem unlikely that *Venturia* forms heterokaryons or has a parasexual cycle in nature. However, Puhalla (1973) recently showed in another pathogen with a similar hyphal organization (*Verticillium dahliae*) that heterokaryosis is restricted to points of anastomosis between strains with complementary auxotrophic markers grown on minimal medium. Hyphae up to 1 mm behind the colony margin are homokaryotic but depend on anastomosis for their growth. At high temperatures, which prevent anastomosis, growth ceases, but it is resumed some days later when prototrophic diploid hyphae develop from some of the heterokaryotic cells formed where anastomosis has occurred at the lower temperature.

The induced mutants of *Venturia* include auxotrophic, morphological, and drug resistant phenotypes, but not changes in pathogenicity on specific varieties. This last class has not yet been sought. The effects of auxotrophy on pathogenicity were studied in detail, but are disappointingly nonspecific and give no clues to the regulation of host specificity (see also page 146). Genetically *Venturia* behaves essentially like *Neurospora* except for some

TABLE 3.4
Segregation for virulence on 4 apple varieties in *Venturia inaequalis*
(Data from Keitt et al. 1948.)

Parent cultures	Haralson	Wealthy	Yellow Transparent	McIntosh
365-2	L*	L	F	F
454-6	F†	F	L	L
No. of asci No. of ascospores				
5 {4	L	L	F	F
{4	F	F	L	L
18 {2	L	L	F	F
{2	F	F	L	L
{2	F	F	F	F
{2	L	L	L	L
12 {4	L	L	L	L
{4	F	F	F	F
35				

*L = Lesion.
†F = Fleck.

inconsistencies in mapping. There are, for example, 10 markers within 10 map units of their centromeres which do not show linkage with each other. This is unexpected in view of the fact that only 7 chromosomes have been observed cytologically. It is also remarkable that of 15 virulence genes tested, all except one are 45 map units or more from their centromeres. Boone and Keitt (1956) also reported 3:1 (6:2) and 1:3 (2:6) aberrant segregation ratios for several markers. One example, involving an albino mutant and its wild-type allele, was easily scored since ascospores carrying the mutant aborted. In a sample of 479 asci from 5 perithecia, 56 showed a 2:6 segregation and 14 a 6:2 segregation of wild type to albino. Several 6:2 asci were dissected, and in each case all 6 normal spores gave rise to wild-type cultures. The overall frequency of aberrant tetrads (14.6 percent), although considerably higher than frequencies of gene conversion found in *Neurospora* or *Sordaria* (.20–.63 percent) (Fincham and Day 1971), is comparable to the higher rates reported in yeast.

The laboratory and greenhouse studies of *Venturia* showed that virulence was simply inherited much like other markers. *Venturia* behaved, in many respects, like the genetically better known *Neurospora crassa*. But *Venturia* was not nearly as convenient an organism as *Neurospora*. Its most important drawbacks were slow growth rate, the length of time needed to make crosses (2 to 3 months), and the need for host material in the greenhouse for studies of virulence. The more sophisticated genetic analysis became, the more the gap widened. Plant pathologists can rarely restrict their analyses to a single genetic background. Strains of different virulence were isolated directly from nature, and genetic analysis was complicated by heterogeneity. Our knowledge of *N. crassa* is largely based on only two wild-type stocks, Abbot and Emerson (Frost 1961), and of *Aspergillus nidulans* on one stock (Dorn 1967). The elimination of heterogeneity greatly simplifies mapping and other studies. The same principle has been followed in many other organisms.

Basidiomycetes: The Rust Fungi

Most rust fungi have two phases, a homokaryon made up of haploid uninucleate cells and a dikaryon made up of binucleate cells. The dikaryon is genetically equivalent to a diploid because each cell of the mycelium normally contains the same two nuclei. Compared with diploids, dikaryons have the added flexibility of being able to exchange nuclei with each other or to donate nuclei to homokaryons to produce new dikaryons. In heterothallic species, each cell of the dikaryon contains two nuclei of different mating type. This imposes some restrictions on nuclear reassortment, since stable recombinants must be heteroallelic for mating type.

The so-called long cycle species have 5 well-defined stages, each with a characteristic spore form. In the microcyclic species several stages are omitted, or at least, have never been found in nature. The five stages are as follows:

Pycnia and Spermatia   Pycnia result from host infection with a basidiospore. They are flask-shaped mycelial structures in the host leaf and produce small uninucleate spermatia (pycniospores), which are carried out through an opening to the leaf surface in a drop of liquid. The pycnia also produce receptive hyphae that protrude through the same opening. A simple mating-type system ensures that "+" spermatia will only fuse with "−" receptive hyphae, or "−" with "+." The transfer of spermatia from one pycnium to another is brought about in nature by insects, but is carried out in the laboratory with an inoculating loop or a small brush. Spermatia cannot initiate infection, and there is no propagation and dissemination of the haplophase, although this is possible in the laboratory by transfer of infected tissue (Hermansen 1959; Patton 1962).

Aecia and Aeciospores   After fusion, the spermatium nucleus passes through the hyphae of the pycnium to an aecial rudiment nearby, which developed at the same time as the pycnium. This migration is accompanied by nuclear division. The rudiment, now dikaryotic, grows to form a pustule (the aecium), which produces chains of binucleate aeciospores. These are discharged and may infect another host plant.

Uredia and Uredospores   The dikaryotic mycelia produced by the aeciospores give rise to sporulating lesions (uredia), which produce binucleate spores called uredospores. Uredospores reinfect other plants, producing further crops of uredospores until conditions are no longer favorable. The uredospores are essentially binucleate conidia and provide a means of clonal propagation for the dikaryon. They can be stored for long periods in liquid nitrogen without much loss of viability (Loegering et al. 1966).

Telia and Teliospores   Aeciospores or uredospores may also give rise to lesions (telia) containing dikaryotic teliospores. Telia are usually produced towards the end of the host's growth period. The teliospores of some rusts are not released, but remain attached to host debris and are a means of overwintering. Teliospores of several long-cycle rusts germinate only after periods of cold and wet, which can be more or less successfully duplicated in the laboratory by freezing and soaking.

Basidiospores   Before germination the two nuclei of the teliospore fuse. At germination the diploid nucleus moves into the germ tube, and there undergoes meiosis. Cross walls delimit four haploid uninucleate cells, each of which forms a basidiospore which is discharged.

In the heteroecious rust fungi, the haploid phase occurs on a different host, often called the alternate host, from that of the dikaryotic phase. For example, in *Puccinia graminis tritici* the aeciospores formed on barberry *(Berberis)* infect wheat or related grasses and cannot reinfect barberry. In autoecious rusts, all phases are produced on the same host. The flax rust fungus *(Melampsora lini)* and the mint rust fungus *(P. menthae)* are examples of autoecious species. In these organisms the genes which determine virulence are expressed in both haploid and dikaryotic phases. Presumably the differences in host range and spore form between haploid and dikaryotic phases of heteroecious rusts are determined by the controlling action of the mating-type gene, which acts as a switch.

AXENIC CULTURE

Hotson and Cutter (1951) described the first successful axenic culture of a rust fungus when they recovered mycelium of the cedar-apple rust fungus *(Gymnosporangium juniperi-virginianae),* growing from infected host callus tissue from a telial gall placed on an agar medium. Although much effort was expended on attempting the culture of this and other rust fungi by similar means, there were few successes (Scott and Maclean 1969). The situation changed dramatically when Williams et al. (1966) described the culture of an Australian race of *P. graminis tritici* (126-ANZ-6,7) from contaminant-free uredospores on a medium containing inorganic salts, sucrose or glucose, yeast extract, and peptone. Further work with other races and other species showed that not all cultures are capable of saprophytic growth, and that even for those that are, not all attempts are successful (Kuhl et al. 1971; Bushnell and Stewart 1971). Uredospores in general will germinate, but frequently fail to develop further. Of those that do initiate some saprophytic growth, only a small fraction form vigorous macroscopic colonies which continue to grow on subculture to a fresh medium.

The most detailed studies, to date, have been carried out with race 126-ANZ-6,7 and are discussed below.

According to Maclean and Scott (1970), macroscopic colonies which appear within two weeks after sowing uredospores, usually sporulate or die within two more weeks and cannot be propagated indefinitely by subculturing. Their mycelia are made up of binucleate cells, and the spores they bear can reinfect wheat. They appear to be normal dikaryons. Williams (1971) showed that the proportion of uredospores which develop in this way depends on the availability of nutrients and can be increased by a 2-hour exposure to 30° C, followed by return to 23° C. It seems that a dikaryotic mycelium is formed only if the germinated spore is able to differentiate a substomatal vesicle. This is a swelling which forms after host infection and penetration through a stomate, and which serves as the origin of the invading dikaryotic mycelium. If development continues on agar without differentiation of a substomatal vesicle, several kinds of abnormal growths appear. One of the commonest has narrow, highly branched hyphae composed of uninucleate cells, and is most likely haploid, arising by breakdown of the dikaryon to its component homokaryons. These "haploid" colonies cannot be propagated by subculturing. Colonies which can be continuously subcultured occur infrequently (about 30 per 10^6 uredospores) and do not appear until four weeks or more after sowing (Maclean and Scott 1970). They form few spores, but even so can be made to infect wheat leaves. Their colonies are made up of uninucleate cells. Although chromosome counts of 6 and 12 suggest that the cells are haploid or diploid, measurements of nuclear volumes in teliospores formed on the host indicate that they are diploid (Maclean: personal communication). Evidently the ability to grow indefinitely in axenic culture is not a property of the majority of uredospores, even of races like 126-ANZ-6,7.

Yaniv and Staples (1972) compared ribosomes from uredospores of *Uromyces phaseoli* (bean rust) and race 126-ANZ-6,7 of *P. graminis tritici* for transferase activity, capacity to bind polyuridylic acid, and leucine incorporation. The spores of bean rust, like those of other rusts which have not been grown in axenic culture, form germ tubes and infection structures within 8 hours when sown on agar media. No further development occurs, and by 24 hours the germlings begin to die. Ribosome preparations of uredospore germlings of both rusts are active during the first 8 hours, but although the activity of 126-ANZ-6,7 continues after 20 hours, that of *U. phaseoli* preparations had declined to a tenth of their former activity.

It looks as if the ability to maintain high rates of ribosome activity through and beyond differentiation of substomatal vesicles is a first requirement for axenic culture. The rarity of forms that can be subcultured continuously suggests that there are other restrictions that are perhaps only overcome by the formation of diploid nuclei.

Developments in the study of axenic cultures of the rust fungi open up possibilities of carrying out genetical and physiological studies independent of the host. They show that *P. graminis tritici* may not always form stable, regularly binucleate dikaryons from uredospores. If crop plants are colonized by similar populations of homokaryotic and heterokaryotic germination products, variation in degree and type of infection will surely follow. As we shall see below, there is no lack of evidence for such variation on the host.

MAKING CROSSES

The method of making sexual crosses in rust fungi is well illustrated by *Melampsora lini* (Flor 1942). To prepare for a cross, teliospores of the two parental races are germinated and the resulting basidiospores used to separately infect susceptible flax plants. In practice, stems bearing overwintered teliospores are suspended over the plants to be inoculated. This may be done in a garbage can lined with wet newspaper where the plants remain for 24 hours. The density of inoculum is varied by exposing the plants for different periods. The object is to obtain well-isolated pycnia arising from single basidiospores. Pycnia formed by two or more compatible basidiospores will produce aecia within a few days and can thus be rejected. Crosses are made by transferring fluid containing spermatia from a pycnium of one race to a pycnium of the other race on another plant. The pycnia are of two mating types, so that only half the crosses are fertile. Fertile crosses form aecia, and the aeciospores from one aecium constitute a single hybrid clone or dikaryon. If a pycnium results from a joint infection by two different basidiospores of the same mating type and is used in a cross, the resulting aecium may be mixed. This is not a problem if each uredospore clone is derived from a single aeciospore per aecium.

Rust cultures may be selfed by pooling pycnial drops and reapplying the mixture to all the pycnia sampled. Again, to avoid the now much greater risk of mixed aecia, single-spore clones (one from each aecium sampled) are taken for the analysis of selfed progeny. These and some other technical problems involved in making rust crosses are discussed by Dinoor et al. (1968a,b) and Miah (1968).

MARKERS

All genetic studies of rusts, to date, have used markers scored during growth on the host plant. Not surprisingly, genes which affect virulence or host range were most commonly employed because of the practical importance of physiologic specialization. In the heteroecious rusts, variation in virulence on the alternate host is known (Green and Johnson 1958; D'Oliveira 1940), but has not been the subject of detailed study. In autoecious rusts, such as *Melampsora lini,* the genes determining virulence would be expected to operate throughout the life cycle. In fact, Flor (1959) showed that this is not always the case. When pycnia on a resistant plant were fertilized with spermatia from an avirulent race, normal aeciospores were formed, but they could not reinfect the host variety on which they were born. Although avirulence was dominant, it was not expressed during nuclear migration and aecial development.

The only other markers that have been used are morphological, the commonest being those that affect spore pigmentation. The spore color mutants of *Puccinia graminis tritici* are a good example. The cytoplasm of the uredospore normally contains an orange carotenoid, and the spore wall is brown. The two colors together produce the reddish brown of wild-type spores. The spontaneous mutant "orange" has a colorless spore wall, whereas another mutant, "grayish brown," has colorless cytoplasm (Newton and Johnson 1927). Both are recessive and when crossed give an F_1 hybrid with wild-type spores. Selfing the F_1 revealed a third class with "white spores" in the F_2. This was doubly mutant, having colorless walls and cytoplasm (Newton et al. 1930). Green (1964) noted that basidiospores from field collections of telia gave rise to white pycnia on barberry in addition to the normal orange pycnia. The color of the aecia formed from white pycnia depended on the source of the spermatia. The cross white × white gave white aeciospores, whereas white × orange gave orange aeciospores. White aeciospores showed poor ability to infect wheat, but the successful infections gave grayish-brown uredospores. Watson and Luig (1962) noted that mutations to colorless wall are common in *P. graminis,* and that all the common Australian races of the stem rust organism are available as orange spore forms as well as wild type.

Many authors have recorded spontaneous and induced mutations affecting virulence and spore color in rust fungi. Some examples are given in Table 3.5; a more complete list can be found in Day (1972a).

SEXUAL RECOMBINATION

Crosses between different races of a number of rust fungi have shown that virulence is generally inherited as a recessive, and that different genes for virulence are usually unlinked and are not allelic. However, some recent work in Adelaide by G. J. Lawrence (personal communication) has shown that multiple allelism or close linkage for virulence does occur in *M.1ini.* Flor (1946) had noted that avirulence on the three flax varieties Akmolinsk, Abyssinian and Leona segregated as a unit. These varieties carry the rust resistance alleles *P1, P2* and *P3* respectively. Lawrence tested the hypothesis that the corresponding determinants for avirulence in the rust *Ap1, Ap2* and *Ap3* were allelic or tightly linked. The allele *Ap* de-

TABLE 3.5
Summary of genetic studies in rust fungi. (Adapted from Day 1971.)

| Rust | Host | Mutation | | | | | Recombination | |
| | | Spontaneous | | Induced | | | | |
		Virulence	Spore color	Virulence	Spore color	Agent	Meiotic	Mitotic
Melampsora lini	flax (*Linum*)		Flor (1965b) (smooth wall)	Flor (1960b) Schwinghamer (1959)		X ray, Fast neutrons, X ray, U.V.	Flor (1956a)	Flor (1957, 1960a, 1964)
Puccinia anomala (*P. hordei*)	barley (*Hordeum*)		Luig & Baker (1956)				d'Oliveira (1940)	
P. coronata	oats (*Avena*)	Zimmer et al. (1963)	Johnson (1949)	Griffiths & Carr (1961)		U.V.	Dinoor et al. (1968a)	Bartos, Fleischmann et al. (1969)
P. graminis avenae	oats				Baker & Teo (1966)	EMS[1]	Green & McKenzie (1967)	Green & Kirmani (1969)
P. g. tritici	wheat (*Triticum*)	Watson & Luig (1968a)	Luig (1967)	Rowell et al. (1963) Luig (1967)	Luig (1967)	X ray, EMS	Kao & Knott (1969)	Watson & Luig (1962) Sharma & Prasada (1969)
P. helianthi	sunflower (*Helianthus*)		Brown (1940)				Craigie (1959)	
P. striiformis	wheat	Macer (1967)		Stubbs (1968)			Miah & Sackston (1970)	Little & Manners (1969 a, b)
P. triticina	wheat		Brown & Johnson (1949)				Brown & Johnson (1949)	
P. recondita	wheat	Samborski (1963)					Samborski & Dyck (1968)	Bartos, Fleischmann et al. (1969)

[1] EMS = ethylmethane sulfonate.

termining avirulence on the variety Koto carrying the resistance allele *P*
was also included in the test. A heterozygous culture of *M.1ini* with the genotype *Ap ap1 ap2 ap3 / ap Ap1 Ap2 Ap3* was prepared and crossed with
a homozygous recessive culture virulent on all 4 host varieties. The
progeny of this test cross were screened by pooling groups of 50–100
aecial pustules and inoculating them to 3 F_1 host lines with the genotypes
P/P1, P/P2 and *P/P3*. The two parental classes *Ap ap1 ap2 ap3* and *ap
Ap1 Ap2 Ap3* are both avirulent on all three screening host genotypes.
Only recombinants such as *ap ap1 ap2 Ap3* etc. are virulent. Three recombinants were found in tests of some 3,160 progenies (the total number of
aecial pustules examined).

Hybrids between the fungi causing wheat stem rust *(P. graminis tritici)*
and rye stem rust *(secalis)* (Johnson 1949), and between those causing oat
stem rust *(avenae)* and *Agrostis* stem rust *(agrostidis)* (Cotter and Roberts
1963), are intermediate in pathogenicity between their parents. They are
frequently avirulent on wheat or oat cultivars which are susceptible to the
common races. This has led to the discovery of resistance genes among
these cultivars that were not known to be present. At the same time, the
hybrids are sometimes found to be virulent on resistant cultivars. Sanghi
and Luig (1971) have pointed to the likelihood that stem rust resistance
transferred to wheat from rye is likely to be overcome by hybrids between
P. graminis tritici and *P. graminis secale.*

However, Green (1971b) examined F_1 and F_2 cultures of such hybrids
and concluded that they were not a threat. All cultures were less virulent
on wheat and rye than either parent. Green suggested that evolution in *P.
graminis* occurred from unspecialized forms, and that the accumulation of
genes for virulence resulted in the appearance of specialized *formae
speciales.* Still more significant is his suggestion that broad combinations
of resistance, derived from interspecific and intergeneric crosses, could
reverse pathogen evolution in the direction of broadened host range at the
expense of reduced destructiveness. In other words, such hybrids would
possess uniform resistance.

Reciprocal crosses in *P. graminis tritici* and in *P. g. avenae* revealed that
virulence to certain differentials is inherited cytoplasmically (Johnson
1946, 1954; Green and McKenzie 1967). The difference between reciprocal
crosses depends on the fact that the spermatial parent contributes little if
any cytoplasm to the hybrid dikaryon. So far, cytoplasmic control of
virulence in the rust fungi has not been demonstrated by heterokaryon
transfer (see page 80). Since the *formae speciales* of *P. graminis* all occur
on barberry, they can be intercrossed.

The observation of segregation for virulence among the progeny of *P.
graminis tritici* races on barberry prompted attempts in the 1930's and

1940's to control rust variation by barberry eradication. The theory was that if no alternate host is available for sexual reproduction, there would be no means for genetic recombination. Large sums of money were spent on destroying barberry plants in the wheat growing areas of North America with doubtful results. In Australia, where barberry is uncommon, *P. graminis* seemed no less able to compete with newly introduced resistant cereals. The reason we now know is that rust fungi have other means of genetic recombination than sexual recombination.

MITOTIC RECOMBINATION

When the uredospores of two different clones of a rust fungus are mixed and inoculated to a host susceptible to both, one may sometimes recover, along with the parent clones, two recombinant dikaryons. The recombinants arise by exchange of partner nuclei. Flor (1964) working with flax rust, and Little and Manners (1969a,b) working with yellow rust (*P. striiformis*), obtained exactly such a result. Not all such experiments are successful; for example, Bartos et al. (1969) reported that no recombinants were recovered from mixed inocula of *P. recondita* (wheat leaf rust fungus). Hyphal anastomosis is required for nuclear exchange. If this were prevented, or nullified, by heterokaryon incompatibility (page 41), no nuclear exchange would be possible. In rather extensive tests with *P. graminis tritici* (Watson and Luig 1958) and *P. coronata* (Bartos et al. 1969), more than two recombinant classes were recovered. It was assumed that genetic recombination involving the transfer of genetic information between different nuclei brought about production of recombinant nuclei. A similar phenomenon was described in matings between homokaryons and dikaryons of the saprophytic Hymenomycete *Schizophyllum commune* (Ellingboe 1965). There is some evidence that recombination occurs in *Schizophyllum* during a transient diploid stage that results from occasional nuclear fusion in the mycelium (Parag 1968). Although a diploid homokaryotic form of *P. graminis tritici* was reported in axenic culture (page 65), there is no evidence that diploids occur on the host.

Hartley and Williams (1971) suggested a simpler theory to explain recovery of more than two recombinants from mixing experiments. They proposed that a dikaryotic culture may contain more than two haploid nuclear genotypes, and that the additional genomes arise by whole chromosome exchange that occurs during synchronous mitotic division. For example, the two haploid nuclei of a dikaryon may be symbolized ABCDEF + A'B'C'D'E'F', where the letters A–F represent six chromosomes and one set of homologues is differentiated from the other by supercripts. Recombinant dikaryons, such as AB'CDEF + A'BC'D'E'F' will, barring position effects, have the same phenotype as the original. This will probably also be true of those with balanced aneuploid nuclei; for example, ABCDEFA'B' + C'D'E'F', and even some with unbalanced combinations provided at least one member of each of the 6 pairs of chromosomes is present, for example ABCD'F + A'B'C'E'F'. The extent of such exchanges in each dikaryon will determine the number of recombinant types recovered from mixing experiments.

The work on *P. coronata* by Bartos et al. (1969) is particularly interesting in that one of the markers under study was unstable. Two races of *P. coronata* were used (393 and 228), which differed in their reactions on 4 oat cultivars (see Table 3.6). Some 215 cultures were recovered from the mixed inoculations; 64 were identified as race 393, 112 as race 228, and the other 39 were either mixtures or nonparental races. Among the latter group, 5 single-spore isolates were identified as race 229 and 4 as race 214. These are probably the classes arising by exchange of nuclei. At the same time two single-spore cultures from the mixture, provisionally identified as race 393 and the nonparental race 229, respectively, were each increased on a rust susceptible cultivar and analyzed further. Twelve single-spore

TABLE 3.6
Reactions of single-spore isolates from two cultures obtained from a mixture of races 228 and 393 of oat crown rust *P. coronata*. (From Bartos et al. 1969.)

Race	No. of isolates	Rust reaction on			
		Victoria	Bond	Ukraine	Saia
Original isolate-race 393 subcultures	–	S	S	R	S
race 228	1	R	R	S	R
229	2	R	R	S	S
296	6	R	S	R	R
216*	1	S	S	S	R
205	2	R	S	R	S
Original isolate-race 229 subcultures	–	R	R	S	S
race 228	4	R	R	S	R
274*	1	S	S	S	R
283	1	S	R	S	R
229	1	R	R	S	S
423	1	S	S	S	S
371	1	S	R	S	S
330	1	R	S	S	S

*Differ in reaction on another variety.

cultures of race 393 showed 5 different virulence combinations, whereas 10 single-spore cultures of race 229 showed 7 combinations. The recombinant phenotypes are shown in Table 3.6. Race 228 was also recovered, once from the first source and four times from the second. Control isolations of the original parental cultures of races 393 and 228 showed them to be stable during the experiments. It was further shown that although the rate of spontaneous mutation to virulence on Saia in race 228 was very high (0.23 percent), the rate with which virulence on Saia was lost in race 229 was very much higher, ranging from 3 in 7 to 18 in 29. In fact, a stable Saia-virulent culture of race 229 could not be obtained. Bartos et al. (1969) showed that cultures of race 229 were made up of regularly binucleate cells, so that its instability could not be explained by the loss of nuclei from cells with 3 or more nuclei. They offered the suggestion that race 229 is an aneuploid and that chromosome loss would account for instability. However, this will only work if virulence on Saia is either dominant or re-quires the presence of a recessive allele for which a null condition cannot be substituted. Another possibility is that virulence on Saia depends on a cytoplasmic element which fails to replicate normally in the presence of a nuclear gene introduced from race 228. If this were so then certain other recombinant races virulent on Saia might be expected to show a similar in-stability. These were not reported.

The reassociation and recombination of nuclei, chromosomes, genes, and cytoplasmic elements accompanying vegetative growth of mixtures of uredospores can evidently accomplish much in the way of generating new forms in some rusts. Indeed, these processes more than make up for the rarity of sexual recombination in stem rust fungi in Australia because of the scarcity of the alternate host, and even extend to exchange between different *formae speciales* (Watson and Luig 1968b). For microcyclic rust fungi such as *P. striiformis*, with no known sexual stage, this is the only means of genetic recombination.

Although recombination mechanisms ensure that a range of genotypes are available for selection, much evidence points to the origin of specific virulence by mutation. Work (noted in Table 3.5) with spontaneous and in-duced virulent mutants of rust fungi has contributed most of our knowledge in this area and is discussed further in chapter 5 (pages 142–145).

Basidiomycetes: The Smut Fungi

Many of the cereal smut fungi are both economically important and easily cultured as haploid cells, so it is not surprising that their genetics has been more thoroughly explored than most groups of pathogens. The general biology of smuts and the earlier genetic studies were covered by Fischer and Holton (1957), whereas Halisky (1965) and Holton et al. (1968) have reviewed much of the more recent work.

The smuts get their name from the sootlike or smutty teliospores (see frontispiece) formed in masses (sori) on leaves, stems, or flower parts of the host. The teliospores are generally unicellular, uninucleate, and diploid. Although they are generally believed to undergo meiosis at germination, there is evidence to suggest that the prophase of meiosis has begun before the spores are released. Thus, like the human egg, the teliospore nucleus may be in an arrested prophase which is completed at germination (P. B. Moens: personal communication). The haploid products, called basidiospores or primary sporidia, are budded off from the germ tube (or promycelium), formed when a teliospore germinates. The haploid forms of many smut fungi can be cultured on simple, defined media as yeastlike cells which multiply by budding.

MATING TYPE

Most smut fungi are heterothallic; that is, their haploid cells are of two or more mating types. Only cells of unlike mating type will fuse and form a pathogenic dikaryon. In all the heterothallic species, cell fusion is controlled by alleles at one locus, and for the great majority, there are only two alleles and consequently only two mating types. There are several possible exceptions. Hoffman and Kendrick (1969) reported that *Tilletia contraversa* (dwarf bunt of wheat) has multiple alleles at a single locus. They found 5 alleles that controlled fusion between haploid cells in 13 collections from the Pacific Northwest. The 5 included 2 that were the same as the two mating types of *T. caries*, which also causes a bunt disease of wheat. A second example, but not so far confirmed by other work, is *Ustilago longissima*, which causes a stripe smut of the grass *Glyceria*. Bauch (1930) claimed to have found 3 alleles at the locus governing cell fusion. Whitehouse (1951) has suggested that the tester stock carrying the "third" allele was a heterokaryon carrying the other two alleles.

In *Ustilago nuda* (loose smut of barley), teliospore germination is followed immediately by dikaryon formation as a result of the fusion of neighboring haploid cells. Homokaryotic haploids can be isolated and cultured by making use of the fact that anthranilic acid (1.25×10^{-3} M) inhibits mitosis but not teliospore germination and meiosis (Nielsen, 1968). On transfer to a medium without anthranilic acid, mitosis takes place and haploid cells may be separated by microsurgery before fusion takes place. The haploids are of either + or − mating type and are mycelial in growth habit. Nielsen (1968) found that all haploid cultures of the + mating type, irrespective of their geographic origin, require an exogenous supply of the amino acid proline for their growth. He suggested that this might be an adaptation favoring rapid formation of the dikaryon, thus promoting the rather unusual type of germination in *U. nuda*. Unlike species which form sporidia at teliospore germination, the dikaryons of *U. nuda* and *U. tritici* are stable in culture. In both species, teliospores are released from the smutted heads of the host in time for them to infect the embryos of developing seeds.

In nearly all of the heterothallic smut fungi, the locus controlling cell fusion is the only mating-type control. However, in *Ustilago maydis*, the cause of boil smut of corn, a second, independent locus controls the stability and pathogenicity of the fusion product. This second locus has multiple alleles. Teliospores are formed only as a result of successful infection and fungus development in the host, and are necessary to complete the sexual cycle. The second locus is thus, in a sense, a mating-type control that works after cell fusion has taken place. The haploid progeny of a dikaryon with the two-locus control are of four mating types, hence the control is "tetrapolar" as distinct from the single-locus, two-allele system which is "bipolar." To be effective in promoting outbreeding, a tetrapolar system must incorporate multiple alleles at both loci. In *U. maydis* only 50 percent of random matings are fertile, since only one of the loci has multiple alleles. The outbreeding efficiency of a bipolar smut is the same; that is, only 50 percent of random matings are compatible. If we compare sib matings, 50 percent are compatible in bipolar forms but only 25 percent are compatible in tetrapolar forms, so that, in theory, inbreeding is reduced in the latter. Of course if tetrapolar species are dispersed and infect their hosts as teliospores rather than as haploid cells, the chances of sib mating and consequent inbreeding are still very high. Bauch's (1930) report, noted earlier, also claims that *U. longissima* is tetrapolar.

Tests for mating type are of two kinds: the first requires host inoculation and is based on the assumption that only paired haploid lines that are compatible will produce disease and teliospores; the second is based either on microscopic observation of fusion among the vegetative cells of two strains or on the macroscopic observation of characteristic dikaryotic infection hyphae from such mixtures. Once a correlation between cell fusion or production of dikaryotic hyphae and pathogenicity has been established, the second method is the method of choice because of its convenience and rapidity. Cell fusion can be scored fairly easily in *U. violacea* (Day and Jones 1968) and several other bipolar smut fungi. However, the production of dikaryotic hyphae, generally known as the Bauch test after its discoverer (Bauch 1932), is even simpler and has been more widely used for both bipolar and tetrapolar forms. It is not without pitfalls. Some isolates will not mate to form dikaryotic hyphae even though they are pathogenic in the host (Puhalla 1968).

The best known species with tetrapolar sexuality is *U. maydis*. The locus controlling cell fusion is designated *a* with two alleles *a1* and *a2*. The *b* locus, with multiple alleles, determines whether the fusion product is pathogenic (Rowell 1955; Rowell and DeVay 1954). Interest in the mating-type controls of *U. maydis* was renewed when Puhalla (1968) showed that the Bauch test could be made to work reliably under the proper conditions. Surveys of the numbers of *b* alleles led to the recovery of 18 from a North American sample (Puhalla 1970) and 12 from a Polish sample (Silva 1972). The two samples totaling 144 cultures had 6 alleles in common. The probable total number of alleles in the combined populations can be estimated at 25.

The germination products of teliospores of *U. maydis* are normally haploid, but diploid cells are sometimes formed through failure of meiosis. Methods for producing diploids at will are also known (Holliday 1961; Puhalla 1969) and have been used in investigating mitotic recombination. Diploids heterozygous for *a* and *b* ($a \neq b \neq$) are pathogenic when inoculated alone (solopathogenic), and give rise to normal teliospores which form haploid products. Puhalla (1968) noted that $a \neq b \neq$ diploids grown under conditions which favor the Bauch test are covered with a mycelium made up of aerial hyphae similar in form to dikaryotic infection hyphae. Diploids homozygous for *b* ($a \neq b =$) were yeastlike and nonpathogenic. This observation led to a very sensitive method for selecting mutants at the *b* locus (Day et al. 1971). Dense platings of $a \neq b =$ diploid cells that had been treated with a mutagen gave occasional mycelial colonies in a confluent background of yeastlike growth, some of which when picked and tested proved to be pathogenic. Haploid *b* mutants (for example, *a1 bDmut*) are solopathogenic and form teliospores but may be forced to participate in crosses if they carry recessive auxotrophic markers such as adenine requirement, which are known to reduce or prevent pathogenicity. In such crosses they are both cross- and self-compatible when mated with haploids carrying different *a* alleles. When mated with their progenitor alleles, the resulting dikaryons show defects which suggest that the *b* locus regulates meiosis and the survival of meiotic products at teliospore germination (Day et al. 1971).

Solopathogenic diploids can be easily detected by their mycelial phenotype when teliospores are plated and allowed to form colonies. A normal colony from a single teliospore is a mosaic of yeastlike cells and dikaryotic hyphae which result from compatible cells that have mated, whereas diploid colonies are entirely covered by mycelium. The frequency of unreduced diploids was 0.26 percent and 0.43 percent in samples of teliospores from two smut galls from a farm crop (Day and Anagnostakis 1971b). Diploids homozygous for *a* ($a = b \neq$) like $a \neq b \neq$ diploids are solopathogenic, but their phenotype varies from yeastlike to mycelial. These diploids cannot be synthesized directly, since cell fusion is blocked by the *a* factor. They are usually recovered as mitotic recombinants from $a \neq b \neq$ diploids and are recognizable by their compatibility with only one *a* tester.

THE NATURE OF THE DIKARYON

For many years mycologists and plant pathologists have assumed that the dikaryon of the smut fungi, like that of other Basidiomycetes, was made up of regularly binucleate cells. Attempts to study the nature of the mycelium in the host met with comparatively little success. In those smuts where seed or seedling infection does not lead to sporulation until quite late in the host's development, the mycelium which keeps pace with host growth is meagre and difficult to detect. The final massive development of mycelia in the host tissue prior to teliospore formation is easy to follow.

The dikaryons of most species could not be studied in culture because they appeared to be obligately parasitic. Most illustrations of hyphal development in the smuts, notably some of the earliest by Brefeld (1883), showed that the tip cells contained cytoplasm but that the bulk of the remaining cells were empty. Day and Anagnostakis (1971a) reported successful culture of the dikaryon of *U. maydis* on a minimal medium containing 1 percent activated charcoal. It was necessary to force the dikaryon by mating two compatible auxotrophic mutant haploid lines with different requirements for growth. Under these conditions haploid parental cells were unable to grow. The presence of charcoal was required because if it was removed by filtration prior to inoculation, the agar medium no longer supported growth of the dikaryon. Whether the charcoal absorbed metabolites which would have allowed cross-feeding between the mutants or whether it absorbed an inhibitor of dikaryon formation was not determined. The form of the dikaryon was identical to the earlier illustrations. The only cells with stainable cytoplasm and nuclei were tip cells and occasional intercalary cells where the branch points were initiated. For the most part such cells were binucleate. Electron microscopy of the mycelium of *U. maydis* shows that the "empty" cells readily collapse during fixation but that they contain membrane structures (Day and Anagnostakis: unpublished).

There seems little reason to doubt that this organization of the mycelial dikaryon is characteristic of most smut fungi that can be cultured. Among other fungi a comparable organization is found in the Entomophthoraceae, a group which includes several insect parasites, a fact also noted by Bauch (1930). One might speculate that this organization, if indeed it is present in the host tissue, represents a reduction of fungal growth to the bare minimum needed to keep up with the development of host tissue. At the same time, the volume of mycelium and the stresss it causes are slight enough that they would not prevent or even delay the host from reaching the stage of development necessary for teliospore formation.

Smut galls from infection by *Ustilago maydis* are formed near the site of inoculation provided the host tissue is meristematic. The dikaryon in this case produces a substance responsible for the neoplastic development of the host tissue to form a gall.

PATHOGENICITY

The pathogenicity of smut fungi is strictly a property of the dikaryon, although we have noted several exceptions such as *U. maydis* diploids and haploids with mutant *b* alleles. The dikaryon has no asexual spore for clonal propagation and infection of other host plants, and is newly synthesized at each infection. Even so, physiologic races of these microorganisms abound (Holton et al. 1968). There is no evidence to suggest that the controls of dikaryon formation determine the varietal host range or virulence of the dikaryon. For the majority of bipolar species, this is not surprising since each dikaryon represents a combination of the same two alleles. For those species with multiple alleles such as *Tilletia contraversa* and *Ustilago maydis*, which attack economic crops, there is little or no evidence on this point since, at least for *U. maydis*, there has been no systematic attempt to define physiologic races. In the case of *U. maydis*, the 25 or so alleles at the *b* locus may well control differences in host range, which further studies will uncover.

Several workers have successfully crossed different species of the smut fungi. Compatibility tests, using the method of observing cell fusion, show that many species and even genera share a common sex control mechanism (Whitehouse 1951). As one might expect, only crosses between fairly closely related forms yield pathogenic dikaryons with fertile teliospores. The studies of Fischer and his students show that such taxonomic characters as host species, sorus type, and teliospore wall markings may segregate like any other markers. Hybrids and segregants with the host range of both parent species are commonly formed in interspecific crosses (see Halisky 1965).

Intraspecific crosses between different physiologic races show that virulence on a given differential host is under simple genetic control, and that virulence is recessive (see page 98). A good example is Sidhu and Person's (1971) study of the genetic control of virulence in *U. hordei*. This species is bipolar. Selfing was carried out by mating the four meiotic products of a single teliospore, isolated by micromanipulation and then propagated as clones, to produce four dikaryotic lines. Crossing was carried out by mating the four haploid lines from one teliospore with the four from another to produce eight dikaryotic lines. Starting with three parental teliospores (D, E, and F), it was shown that all three produced selfed or crossed dikaryons that could be scored as virulent or avirulent on three barley varieties. Virulent dikaryons produced one or more smutted spikes on about 50 percent of plants raised from inoculated seeds. Avirulent dikaryons produced no smutted heads. The four selfed lines from D were virulent on Excelsior but avirulent on Hannchen and Vantage. The selfed lines from spores E and F, however, were avirulent on Excelsior but virulent on Hannchen and Vantage. All dikaryons from the crosses D × E and D × F were avirulent on Excelsior, Hannchen, and Vantage, showing that virulence on all three is recessive. The progenies of 3 hybrid teliospores from each cross were then backcrossed to E and gave segregations showing that they were each heterozygous at a single locus (*Uh-1*) controlling virulence on Hannchen and Vantage. Backcrosses to D showed that virulence on Excelsior was controlled by another single locus (*Uh-2*). Earlier studies by Thomas and Person (1965) had shown that the difference between very low levels of virulence (about 5 percent infection) and avirulence on the barley varieties Gateway and Olli was also controlled by alleles at a single locus. The loci *Uh-1* and *Uh-2* control higher levels of virulence that range from 11 to 86 percent (mean 49.7 percent).

MITOTIC RECOMBINATION

The regular occurrence of meiosis at teliospore germination suggests that mitotic recombination would be of little or no adaptive significance in the smut fungi. Its usefulness as a tool in mapping and genetic studies of several species has been amply demonstrated. The spontaneous frequency of mitotic recombination in diploids of *U. maydis* and *U. violacea* is very low but can be increased by ultraviolet light, and in *U. maydis*, by drugs like mitomycin and fluorodeoxyuridine, which inhibit DNA synthesis or repair (Esposito and Holliday 1964; Holliday 1964). Haploidization by mitotic nondisjunction seems not to occur in *U. maydis* even after treatment with parafluorophenylalanine, which induces it in *U. violacea* (Day and Jones 1969).

Although demonstration of these phenomena in the laboratory is of considerable interest, we may ask whether they indeed occur in the host and in the field. Kozar (1969), using barley plants in a glasshouse, showed that mitotic recombination can occur in *U. hordei* dikaryons carrying induced markers. There is also evidence to suggest that it occurs in field infections of *U. maydis*. Day and Anagnostakis (1971b) noted that some colonies produced by single teliospores from field infections were not mosaic or covered with mycelium like those described earlier (page 75), but were entirely yeast-like. A proportion of these were made up of haploid cells, either of only one mating type or of two mating types that were incompatible and were derived from incomplete tetrads in which at least two products had failed to survive. Some others were $a \neq b$ = diploids, and it was suggested they arose by mitotic recombination just prior to teliospore formation. An important feature of the work is that it used markers carried by all natural infections of *U. maydis* and showed that they are as useful as induced mutants for genetic analysis.

INTERSTRAIN INTERACTIONS

While testing a collection of North American isolates of *U. maydis* for mating type, Puhalla (1968) noticed that the Bauch test between certain isolates failed, although the isolates were compatible when inoculated to corn seedlings. One class of isolate, called Pl, produced a proteinaceous substance which diffused from the colony, killing nearby cells of the other class (Hankin and Puhalla 1971). This second, sensitive class was called P2. Analysis of the progeny of crosses between P1 and P2 showed that the difference was cytoplasmically determined. *U. maydis* does not show maternal inheritance, where one partner of a cross contributes the bulk of the cytoplasm to the progeny. The evidence for cytoplasmic inheritance lay in the fact that P1 and P2 did not segregate in tetrads. From 90 to 99 percent of the tetrads gave rise only to P1 cells with normal segregation of other markers. Some of the remaining tetrads were wholly P2, whereas the rest were wholly of another class called P3. Both P2 and P3 tetrads showed normal segregation of other markers. The P3 cells neither formed the killer material nor were sensitive to it. Puhalla suggested that P1 cells contained two kinds of extranuclear particles: [I] responsible for killer substance and [S] responsible for resistance to the killer substance. The P2 cells contained neither, whereas the P3 cells contained only [S]. The evidence for the nature of P3 came from crosses of P2 × P3, which showed that P3 was controlled by an extranuclear determinant [S]. A further complication is that P2 cells carry one of two alleles of a chromosomal gene: s determines sensitivity to killer, whereas s^+ determines resistance. In diploid P2 cells, sensitivity (s) is dominant. These relationships are summarized in Table 3.7. Support for Puhalla's explanation followed when it was shown that P1 and P2s cells would mate on minimal agar medium containing 1 percent activated charcoal. When dikaryotic hyphae from such matings were allowed to dissociate on media without charcoal, the cells which carried the mating type alleles associated with the P2s parent were found to be P1 in phenotype. Similarly, by isolating the dikaryon from the cross P3 × P2s and allowing it to break down, only P3 cells were recovered, but again they were of the two classes corresponding to the

TABLE 3.7
Forms of *Ustilago maydis* defined by interstrain interactions. (Sources given in text.)

Character	Genotype	Phenotype	VLPs
Killer	s or s^+ [I] [S]	P1 produces killer protein, insensitive	+
Sensitive	s	P2s, sensitive	−
Neutral	s^+	P2s^+ no killer, insensitive	−
	s or s^+ [S]	P3 no killer, insensitive	+

parental mating types (Day and Anagnostakis 1973). One assumes that in each experiment, extranuclear particles, contributed to the dikaryon by one parent, were distributed to the breakdown products.

Both P1 and P3 cells contain viruslike particles 41 nm in diameter that are transmitted at dikaryon formation (Wood and Bozarth 1973). Both cell types contain virus particles that are indistinguishable by all physical tests applied so far. For the time being, only the resistance factor [S] can be associated with the virus. One possibility is that [I] and [S] are forms of the same virus, and that [S] differs from [I] only by its inability to direct synthesis of killer protein. Transfer of virus infection through hetero-karyosis was first used by Lhoas (1971a) to transmit viruslike particles between strains of *Penicillium chrysogenum*. Lhoas (1971b) later showed that naked fungal protoplasts could also take up the particles and develop into "infected" mycelium.

Several other interactions between strains of *U. maydis* are known. Puhalla recorded two other inhibitory conditions like P1 which differed from it in their specificity for sensitive clones. A different kind of interaction, the so-called browning reaction, in which sensitive cells be-came darker and produced thicker walls as a result of contact with another strain, was also shown to be controlled in part by extranuclear de-terminants (Anagnostakis 1971). Another example of induced mor-phological change, discussed by Silva (1972), has not been studied further.

The implications of killer-type interactions for controlling pathogens are discussed later on page 88.

Fungi Imperfecti

A large number of plant pathogenic fungi are known only from conidial stages. They have no known sexual stages and are grouped in the imperfect fungi. Within this group they are classified according to their spore and hyphal morphology. Many have affinities with the Ascomycetes. For some, a perfect, or ascus producing, stage was later found and given another generic name, so that the organism is then known by two names (for example, *Helminthosporium sativum* and *Cochliobolus sativus*).

Practical experience with imperfect pathogens early showed that breeding for disease resistance resulted in physiologic specialization. The first account of physiologic races was by Barrus (1911), who described three races of the imperfect fungus *Colletotrichum lindemuthianum* causing bean anthracnose (dark, sunken spots), which were distinguished by different bean cultivars.

The imperfect fungal pathogens include those that cause wilts *(Fusarium, Verticillium)*, leaf diseases *(Ascochyta, Piricularia),* and storage rots *(Penicillium, Aspergillus).* In all these genera, parasexual recombination has been demonstrated in laboratory studies with induced markers (Tinline and MacNeill 1969). Eric Buxton was one of the first plant pathologists to appreciate the importance of parasexual recombination, and his work on *Fusarium oxysporum* (Buxton 1956) provided a model for studies of other genera and species. The method, developed by Pontecorvo (1956) with the saprophyte *Aspergillus nidulans,* was to isolate auxotrophs following the treatment of spores with a mutagen and to use them, in complementary pairs, to synthesize heterokaryons which could grow on a minimal medium. The heterokaryon formed homokaryotic spores and these were spread on a minimal medium. Any diploid spores derived from chance fusion of two complementary nuclei of the heterokaryon would grow. The remaining homokaryotic haploid spores would not grow, providing the auxotrophic markers were stringent and no cross-feeding by diffusion of metabolites occurred. Although such diploids are rare, selection is so efficient that their recovery is not usually a problem. Heterozygous diploids may undergo mitotic crossing-over or become haploid by random loss of chromosomes at mitosis (nondisjunction). Either mechanism results in the appearance of new phenotypes.

The application of this method to a number of other fungi showed that they too could be made to behave in the same way. In *Aspergillus nidulans* the sexual cycle was some 500 times more active than the parasexual cycle in forming recombinant genotypes. However, it seemed that in the imperfect fungi the rates of variation in the parasexual cycle could be rather higher. For example, Lhoas (1967) found that rates of mitotic recombination in the imperfect *Aspergillus niger* were some ten times higher than in *A. nidulans*. Hastie (1970a) found that, compared with *A. nidulans,* heterokaryons and diploid cultures of *Verticillium alboatrum* are very unstable. For example, the frequency of mitotic recombination per nuclear division in the phialides (spore bearing structures) of *Verticillium* diploids, affecting some part of the genome, was estimated to be at least 0.2. Käfer (1961) estimated the frequency of mitotic crossing-over, or nondisjunction, at 0.04 per mitosis in *A. nidulans*. Of course rates for individual loci are much lower. No figures are available for *Verticillium*, but in diploid yeast Thornton and Johnston (1971) recently determined spontaneous rates of mitotic recombination at 4 different loci. They ranged from 3.5×10^{-4} down to less than 2×10^{-6}.

I have chosen some recent work on *Piricularia oryzae* to illustrate some of these points. Although some other pathogens are more fully understood with regard to parasexuality, *Piricularia* is particularly interesting for the light it sheds on variation in pathogenicity.

RICE BLAST

Piricularia oryzae is an important pathogen of rice *(Oryza sativa)* that causes a disease of leaves, nodes, and inflorescence called blast. Low night temperatures prior to inoculation (15–20° C) favor lesion development (Sadasivan et al. 1965), but optimum temperature for infection and disease development is 24–26° C. Infection occurs by formation of an appressorium followed by direct penetration of the host epidermis. Mycelial growth in the host is accompanied by toxin production, which causes host cell death ahead of the mycelium (Iwasaki et al. 1969). Within a week of infection, conidia are formed by aerial conidiophores, an average sized lesion producing 4,000–6,000 conidia per night for two weeks (Ou and Ayad 1968). The conidia are pyriform generally with 3 cells.

Variation in Culture   In one of the most comprehensive accounts of variation in *Piricularia*, Yamasaki and Niizeki (1965) reported that in 28 different strains of *P. oryzae*, the majority of cells in conidiophores, conidia, germ tubes, and mycelia were uninucleate. The same was also true of a number of other strains of *Piricularia* from other hosts. The frequencies of cells with 2 or more nuclei ranged from 1.1 to 4.6 percent except for a group of related isolates with 11.6 to 16.8 percent which included Saka-2 mentioned below. Yamasaki and Niizeki included 3 cultures of *P. oryzae* examined by Suzuki (1965), who claimed the cells had an average of 4 or 5 nuclei each with a range of 1 to 13. They also noted that the nuclei of a conidium are all derived from the single nucleus in the conidiophore.

They then examined a total of 295 different isolates of *Piricularia*, all of single-spore origin, for cultural variation. Each culture was grown on 5 dishes of potato sucrose agar, incubated, and then scored for morphological variants appearing as sectors in the colonies. Some 82 of the isolates were from hosts other than rice; the remainder were presumably all *P. oryzae*. Only 50 isolates (17 percent) formed sectors, several doing so at a high frequency. Three of this latter class were studied further. One, Saka-2, continued to vary during 9 further generations of consecutive subculture. The variants that arose in the 3 isolates were examined to see whether a high frequency of cells with 2 or more nuclei was correlated with instability. It did not appear to be. Some of the data are shown in Table 3.8. Isolate 56, only a little less variable than Saka-2, did not differ significantly in percentage of multinucleate cells from isolates that were stable. The table also shows that in Saka-2 the number of nuclei per cell is itself quite variable.

In their search for an explanation of this variation, Yamasaki and Niizeki showed that hyphal anastomosis occurred readily between certain strains and was followed by movement of nuclei which, in some instances, appeared to lead to their fusion. They synthesized heterokaryons from paired auxotrophic mutants induced in several strains of *P. oryzae,* some of which carried other markers. Resistance to copper sulphate behaved as a dominant, and production of hydrogen sulphide as a recessive. Single-spore cultures from 4 of these heterokaryons were examined. Two examples are shown in Table 3.9. All 4 heterokaryons generated recombinant phenotypes. Heterokaryon I gave rise to markers not seen in either of its parents. The data are most simply explained by a parasexual mechanism of a type where the diploid is very unstable and shows continued segregation either by haploidization (through nondisjunction), or by mitotic crossing-over, or by both mechanisms. Hastie (1970a) described a

TABLE 3.8
Numbers of nuclei in mycelial cells of the strains Saka-2 and 56 of *Piricularia oryzae* and in variants which appeared as sectors. Strains X and 41 represent the extremes of 27 variants from Saka-2. Strains 10 and 16 represent extremes of 8 variants from 56. (7), (8), (9) indicate consecutive subcultures on potato sucrose agar. (Data from Yamasaki and Niizeki 1965.)

| Strain | No. of nuclei in mycelial cells | | | | | | No. of cells observed | % cells with 2 or more nuclei | Average no. of nuclei per cell |
	1	2	3	4	5	6			
Saka-2	509	61	13	4			587	3.2	1.17
⁻X (7)	401	105	9	10	1	1	527	23.9	1.31
(8)	500	64					564	11.3	1.11
(9)	478	77	1				556	14.0	1.14
⁻41 (7)	522	10					532	1.8	1.02
(9)	574	34	2				610	5.9	1.13
56	520	13					533	2.4	1.02
⁻10 (7)	518	20					538	3.7	1.07
⁻15 (7)	551	16					567	2.8	1.06
⁻16 (7)	511	8	1				520	1.7	1.03

TABLE 3.9

Phenotypes of parents, heterokaryons and derived cultures of *Piricularia oryzae*. (Data from Yamasaki and Niizeki 1965.)

	Auxotrophy						Colony color	No. of cultures
	Ad.	Meth.	Inos.	Lys.	H$_2$S	CuSO$_4$		
Parent 1	−	+	+	+	−	S	black	
Parent 2	+	+	−	+	−	R	black	
Heterokaryon I	+	+	+	+	−	R	black	
Single-spore cultures	+	+	−	+	−	R	dark-gray	31
	+	−	+	+	−	S	black	110
	+	+	+	+	−	R	brown	15 (Het. I)
	+	+	−	+	−	S	black-brown	2
	+	+	−	+	−	R	gray	1
	−	+	+	+	−	S	gray	1
Total								160
Parent 3	+	−	−	+	+	S	black	
Parent 4	−	+	+	−	−	S	black	
Heterokaryon IV	+	+	+	+	−	S	black	
Single-spore cultures	+	−	−	+	+	S	black	46 (Par. 3)
	−	+	+	−	−	S	black	26 (Par. 4)
	+	+	−	+	+	S	dark-gray	65
	+	−	+	+	+	S	black	2
	+	−	−	+	+	S	black-brown	10
Total								149

situation of exactly this kind in *Verticillium alboatrum* (see page 81). Yamasaki and Niizeki observed mitotic divisions in mycelia and conidia of wild-type isolates, and suggested chromosome counts of 3 and 5 or 6. Certainly the behavior both of wild-type isolates such as Saka-2 and 56 and of the synthetic heterokaryons is readily explained if they carry unstable heterozygous diploid nuclei.

A later account of variation in *Piricularia* by Suzuki (1967), which did not make use of induced markers, confirmed many of these observations. The chief difference lay in the interpretation of cytological details. This is not a problem peculiar to *Piricularia*. Suzuki claimed that most cells were multinucleate, and that this led to what he called "persistent heterokaryosis." Several reports have claimed to resolve the question of the number of nuclei per cell by electron microscopy, but they did not deal with many cells, examined in serial sections, and so are not definitive.

Variation on the Host The control of rice blast by breeding for resistance began in India and Japan in the 1920's (Ou and Jennings 1969). Oligogenic race-specific resistance has been employed with the familiar consequence of physiologic specialization. There are many varieties with general resistance which may vary with environmental conditions. Physiologic races of blast were first recognized in 1922 in Japan, but were not considered in breeding programs until 1960 (Takahashi 1965). In recent years a succession of varieties has been introduced in the major rice growing areas as established varieties became susceptible to new races and were replaced. The extent of pathogenic variability is the subject of some debate. Ou and Ayad (1968) claimed to have recovered 22 different physiologic races among 56 single-conidial cultures from one typical field lesion on a susceptible rice cultivar in the Philippines. Some 14 races were found among 44 cultures derived from a second lesion taken from the same field a month later. Further single-conidial cultures derived from several of the original single-conidial cultures showed a comparable range of races: 9 and 10 races each from two such sets of 25 cultures. The predominant races derived from the two lesions, collected a month apart, were also different. The authors commented, ". . . if more varieties were used as differentials, each conidium in the culture could probably be shown to be a distinct race." In fact, only one of the 17 differentials used in the study remained consistently resistant and only one susceptible. Some resistant differentials had only one or two lesions on very few leaves. Reisolation from these few lesions, followed by reinoculation, resulted in greatly increased numbers of lesions. Evidently selection brought about an increase in the average number of virulent conidia when cultures were cycled on such plants.

Giatgong and Fredericksen (1969) also reported variability during three successive generations of single-spore cultures of 2 races during pathogenicity tests on 4 rice differentials that were carried out in Texas. Each generation was represented by 20 single-spore cultures. The data obtained for one of the races is shown in Table 3.10. The parental class was the most frequent in each generation. Even so, 9 of the 16 classes recognizable with 4 differentials were found among the 60 isolates.

In direct contrast Latterell (1972) found that changes in race pattern were rare in tests involving 600 single-spore isolates from cultures and lesions. The discrepancy probably arises chiefly through differences in interpretation of variation in lesion characteristics on a single leaf that may be due to physiological or developmental differences between different parts of the leaf or to differences in micro-climate. The use of different inoculation methods by different workers is another source of discrepancy. In this connection Manibhushanrao and Day (1972) described differences in rice cultivar response to temperature treatments prior to inoculation that would profoundly affect race identification. A third possibility is the

TABLE 3.10
Disease reactions shown by 20 single-spore cultures of *Piricularia oryzae* (race 1) in 3 successive generations. (Data from Giatgong and Frederiksen 1969.)

Rice differentials	Parent isolate	Generations and reaction classes														
		1				2							3			
Zenith	S	S	S	S	R*	R	R	S	S	R	S	R*	R	R	R	R
CI 8970-P	R	R	S	R	S	S	R	R	S	R	S	S	S	S	R	S
CI 8970-S	S	S	S	R	S	S	S	S	S	R	S	S	S	S	R	R
PI 180061	S	S	S	S	R	R	R	S	S	R	R	S	S	R	R	R
	FREQUENCY	13	4	2	1	7	5	2	1	1	1	3	15	2	2	1

*Reaction of isolate used as source of next generation.

high risk of contamination by spores of other races when tests are carried out near rice crops. This can be gauged by using uninoculated check plants. As Latterell pointed out this debate must be quickly resolved so that the value of breeding for specific resistance can be clearly assessed.

Unfortunately, the genetic basis of resistance in most of the rice differentials is still unknown. For this reason it may be misleading to compare the *Piricularia* races with races of other pathogenic fungi. One could argue that much of the variation observed by Ou and Ayad, and used by them to define races, is under minor gene control. This kind of variation, although it may be present, is often ignored in other better categorized systems.

CYTOPLASMIC INHERITANCE AND MYCOVIRUSES

Until a few years ago a discussion of cytoplasmic inheritance in plant pathogenic fungi would have been complete with an account of Johnson's (1946) discovery that virulence of the stem rust fungus *(Puccinia graminis tritici)* to the wheat cultivar Marquis is inherited cytoplasmically. The observation depended on the fact that the pycnial parent (see page 66) contributes the bulk of the cytoplasm to a rust dikaryon, and that a phenotypic difference between reciprocally constituted dikaryons indicates that the cytoplasm is responsible.

A corollary of a reciprocal difference between crosses is that no further segregation is expected when either kind of zygote undergoes meiosis. Puhalla (1968) used this last feature to show that in the smut fungus *Ustilago maydis,* "killer" protein production (P1) and one form of resistance to killer (P3) are both cytoplasmically inherited. I have already reviewed (page 80) the confirmatory evidence showing that these properties may be transferred by making heterokaryons between "killer" (or P3) and sensitive cells, and that P3 (if not "killer") may be due to a virus. In recent years virus infections of fungi (mycoviruses) have been found to be fairly common, and the following questions arise: To what extent may other examples of cytoplasmic inheritance in fungi be explained by virus infection? Can mycoviruses be used to control plant pathogenic fungi? How do mycoviruses affect the host-parasite interaction?

Much of the recent literature on mycoviruses was recently reviewed by Hollings and Stone (1971) and Bozarth (1972), who also posed these questions and put forward some candidate examples and some preliminary results. Table 3.11 lists some plant pathogens where there is evidence of cytoplasmic inheritance from sexual analysis, or heterokaryon tests, and some examples which are still not completely understood. In some of these, and in some other examples for which there is no genetic information, viruslike particles (VLPs) have been identified by electron microscopy. An examination of these examples will help to find partial answers to the last two questions.

The majority of examples of VLPs in the fungi were found in surveys of culture collections where there was usually no reason for suspecting their presence. For example, Richards (1971) reported that examination of some 350 isolates of fungi belonging to most of the major groups revealed that 44, or 12 percent, carried VLPs. Proving that any particular instance of cytoplasmic inheritance is determined by a mycovirus will be quite

TABLE 3.11
Cytoplasmic inheritance and incidence of VLPs in plant pathogenic fungi.

Fungus	Host	Marker	VLPs	Reference
Endothia parasitica	Chestnut	hypovirulence	?	Grente & Sauret (1969) Grente (1971)
H. maydis	Corn	—	+	Bozarth et al. (1972)
H. oryzae	Rice	"disease"	?	Lindberg (1971)
Helminthosporium victoriae	Oats	"disease"	+*	Lindberg (1959, 1968, 1969) Lindberg & Pirone (1963)
Ophiobolus graminis	Wheat	hypovirulence	+	Lapierre et al. (1970) Lemaire et al. (1971)
Pestalozzia annulata	Cinchona	"infectious mutant"	?	Chevaugeon & Lefort (1960) Chevaugeon (1968)
Piricularia oryzae	Rice	"lysis"	+	Férault et al. (1971)
Puccinia graminis tritici	Wheat	Virulence	?	Johnson (1946, 1954)
Puccinia graminis avenae	Oats	Virulence	?	Green & McKenzie (1967)
Ustilago maydis	Corn	Killer (P1)	+	Puhalla (1968) Day & Anagnostakis (1973) Wood & Bozarth (1973)

*Bozarth (pers. commun.)

difficult. Since VLPs seem to be fairly common, an association between VLPs and a particular phenotype could be fortuitous. The problem is further complicated by the difficulty of infecting intact mycelia with either crude or purified VLPs. Lhoas (1971b) infected naked protoplasts of *Penicillium stoloniferum,* prepared by removing wall material with chitinase, by suspending them in a cell-free preparation of VLPs.

There is clearly little hope at the moment of using suspensions of pathogenic mycoviruses to control plant disease fungi ùnless a means can be found of promoting their uptake, perhaps through newly formed cell walls or through walls that have been broken down or weakened in some way as happens prior to cell fusion. For the time being, anastomosis appears to be the most useful method of obtaining mycovirus transmission. Although heterokaryon incompatibility would seem to prevent or interfere with the transmission of mycoviruses and other cytoplasmic determinants, it may not be very efficient as the following example will show.

Hypovirulence in the chestnut blight fungus *(Endothia parasitica)* is a condition of reduced aggressiveness that enables the chestnut host to effectively arrest the pathogen and prevent its further spread. Hypovirulence appears to be cytoplasmically inherited. Grente and Sauret (1969) compared simultaneous and consecutive inoculations of virulent and hypovirulent strains of various origins to see whether hypovirulence could be used to control infection. Simultaneous inoculation appeared to be the more stringent test, since only 28 percent of inoculations gave rise to arrested cankers within 11 to 13 months. By far the most effective combination (96 percent successful) was one in which the hypovirulent and virulent strains were probably related since they were of the same geographical origin. Where the hypovirulent strain was applied 8 months after the first virulent inoculum, control was achieved in 60 percent of the cases 7 months later. Laboratory studies confirmed that with certain pairs of stocks, cell death followed anastomosis. This symptom of heterokaryon incompatibility would undoubtedly restrict infection by an invasive cytoplasmic element. The reason why such combinations were effective in consecutive inoculation could have been a lessening of the heterokaryon incompatibility reaction in older mycelia established in host tissue.

Caten (1972) has examined the role of heterokaryon incompatibility in restricting transfer of the highly suppressive cytoplasmically inherited condition vegetative death in *Aspergillus amstelodamii.* Transfer was 100% effective between compatible isolates, 15% effective between isolates whose incompatibility was determined by a single gene, and completely prevented where more than one gene determined incompatibility. Vegetative death shows an abnormal cytochrome spectrum and seems likely to be a mitochondrial abnormality like the *poky* mutant of *Neurospora crassa* rather than a mycovirus (Caten-personal communication).

The killer phenomenon in *Ustilago* appears to be rather unpromising as a control agent. Resistance to killer, determined by a nuclear gene, is more common than susceptibility. In host tissue, killer cells do not kill sensitive cells, but if mating can occur, they convert them to killers. However, more useful killer forms that are toxic to more strains under a greater range of conditions may be discovered or selected.

A virus is associated with a loss of pathogenicity in the cereal pathogen *Ophiobolus graminis* (Lapierre et al. 1970). A virus-infected culture of *Ophiobolus* was shown to bring about partial control of take-all disease of wheat seedlings when added to soil containing the normally virulent uninfected form of the pathogen (Lemaire et al. 1971).

The diseases of *Helminthosporium* are excellent candidates for virus infections but so far have not been examined for VLPs. One, a disease of *H. oryzae*, causes the fungus to produce a toxin which stunts the growth of rice seedling roots. The infected strain also causes a more severe disease than the typical uninfected *H. oryzae* (Lindberg 1971). Evidently infections, whether of viruses or some other invasive cytoplasmic element, can have profound effects on the host-parasite interaction, and these effects may either increase or decrease pathogenicity.

NEMATODES

Until recent years the genetics of plant pathogenic nematodes was neglected because plant pathologists concentrated on describing species and developing control methods based on soil fumigation, soil drenches, or systemic nematicides. As resistant varieties of crop plants began to be used, physiologic races appeared and nematode biotypes were described that were differentiated only by their reactions on resistant host varieties. One of the best known species showing this behavior is the potato cyst-nematode or golden nematode *(Heterodera rostochiensis)*. In this species, only those larvae which successfully stimulate the host root to produce the giant cells on which they feed will reproduce. Giant cells are formed in response to the injection of larval saliva. The larvae are hermaphroditic, but once established in the host root they become female, producing eggs after mating. Larvae which do not induce giant cell formation either become males or die. To make controlled matings, Webster (1965) devised a method for separating male from female larvae. Potato roots growing in sand were bathed in a suspension of newly hatched larvae. After 18 days the sand was replaced by an aerated nutrient solution changed every 2 days. Under these conditions, when the males mature, they fall to the bottom of the vessel containing the nutrient solution and do not fertilize the females which remain attached to the roots. The collected males were stored at 4°C, and when no additional males were produced (52 days), the roots were covered again with sand and a suspension of males of known origin added. Two weeks later, dark brown cysts full of eggs were found on roots with added males. No eggs were found on roots that had not received males, showing that parthenogenesis did not occur. Parrott (1968) has since made single pair matings on agar.

The first attempts to establish the genetic control of host range in *H. rostochiensis* showed that European races belong to two species (Jones and Parrott 1965). Table 3.12 shows the races recognized in Britain. A number of other races are recognized elsewhere in Europe (Ross and Huijsman 1969) but need not concern us here. Jones et al. (1970) made crosses between individual males and females of the three races in all possible combinations, scoring the percentages of females that produced eggs. Their figures indicate the fertility of the crosses and are shown in Table 3.13.

The matings A × B and A × E were less fertile than B × E or those between nematodes of the same race. Further examination showed that the larvae of race A are shorter than those of B and E and have shorter stylets. Females of A are yellow, whereas females of B and E are cream and white, respectively. It appears that race A belongs to the species *H. rostochiensis,* that the two other races belong to a separate undescribed species, and that physiologic races exist in both species. At the present time the taxonomy is therefore rather confused, and it is not surprising that studies of the inheritance of pathogenicity were confounded by low fertility.

Races are known in several other species of nematodes also (Sturhan 1971). Some, like the more than 20 races of the stem nematode *Ditylenchus dipsaci,* are differentiated by host species belonging to many different plant families. However, the biotypes of this species appear to belong to a common breeding continuum (Webster 1967), a situation that may parallel that in *Helminthosporium.*

The current interest in nematodes as a tool for studying the genetics of a simple nervous system (Brenner: in preparation) has important implica-

TABLE 3.12
British races of *Heterodera rostochiensis* defined by two genes for resistance.

Race	H_1 *S. tuberosum andigena*	H_2 *S. multidissectum*
A	−	+
B	+	−
E	+	+

TABLE 3.13
The percentages of females with eggs from
all possible single pair matings between 3
races of *H. rostochiensis*. (From Jones et al.
1970.)

		Females		
		A	B	E
	A	61	0	15
Males	B	39	60	60
	E	38	68	67

Note: From 50 to 250 single-pair matings were
attempted in each category.

tions for work on host-parasite relationships. For example, studies with
the nonparasitic species *Caenorhabditis elegans* and *Ascaris lumbricoides*
have developed methods for recovering induced morphological, behavioral, and drug-resistant mutants (Ward 1973). Similar markers would
be very useful in genetic studies of plant parasitic forms, and some of them
may help in investigating the genetic controls of the behavior patterns in
host-seeking, feeding, and reproduction.

It is probably no accident that the nematodes that show a high degree of
host specificity are endoparasitic. They reproduce in close association with
their host tissue and are subject to stringent selection to maximize compatibility, so as not to prejudice reproduction.

In recent years knowledge of the genetic systems of plant parasitic
nematodes has greatly increased. In a recent review, Triantaphyllou (1971)
listed the chromosome numbers of more than 70 species together with information on their modes of reproduction. There are surely many techniques and methods for genetic study awaiting discovery and application in
this important group of parasites.

4

THE GENE-FOR-GENE CONCEPT

If host and parasite are both to survive during evolution, we can expect that genetic changes in one will be balanced by changes in the other, which will tend to restore equilibrium. The directions of such changes are necessarily opposed—toward greater resistance in the host and greater virulence in the parasite. The converse—change to lowered resistance in the host and to lowered virulence in the parasite—may well increase fitness in both and is not so unlikely as it may seem. I will discuss some examples in Chapter 7. Making a simple statement of this interdependence required equally extensive knowledge of both host and parasite. The knowledge of the genetic control of resistance grew from the practical operation of breeding new resistant varieties of crop plants. Knowledge of the parasite had no such immediate practical rewards, but until it was available, little progress in understanding the interaction was possible. This was a pursuit for basic science.

THE EMERGENCE OF A FORMAL GENETIC RELATIONSHIP

As the extent of physiologic specialization of the cereal rusts became apparent, pathologists began to search for order in the chaos of innumerable races. Writing in 1935 Vavilov said:

> Theoretically, it is very probable that the number of physiologic races of parasites is much greater than has been determined up to this time. The closer we get to the sources of origin of species and forms of cultivated plants, and closely related wild species, the more probable it is that we will find new races of parasites corresponding to these (Vavilov 1949).

This was an accurate forecast of what was to be found nearly 20 years later when the races of *Phytophthora infestans* in central Mexico were examined. They occurred in great variety on wild potato species used extensively as sources of blight resistance in breeding programs (Niederhauser 1956). Similar discoveries were to be made for powdery mildew (*Erysiphe graminis*) on cereals in Israel (Koltin et al. 1963).

At first, races were designated by numbers, or letters, assigned to them as they were discovered. The host range of the numbered races was recorded in tables showing reactions on differential hosts. A variety of problems followed: How many classes of reactions should be taken into account—resistance versus susceptibility or intermediate classes as well? How many differentials are needed? As more were used, the labor of race identification increased. As these problems were tackled, some plant breeders decided to designate races by formulae which indicated host range, making it unnecessary to use a reaction table. Sets of differential hosts were altered and simplified to include only varieties needed to identify races that were dangerous or potentially so. Most interactions were scored R or S for race identification, and some included X for a mixed, or mesothetic, reaction which has both resistant and susceptible lesions. The genetic basis of resistance in the differential host was known in many cases. Although knowledge of the inheritance of pathogenicity showed that it could also be under simple genetic control, no one until 1942 had completed genetic studies of both partners.

PROOF

In 1942 Flor published an account of segregation in the progenies of crosses between different races of the rust fungus *Melampsora lini* for virulence on flax. The flax varieties Bombay and Akmolinsk were known to carry single independent genes for resistance to the races used in the

TABLE 4.1
Inheritance of resistance to races 22 and 24 of *Melampsora lini* in two
varieties of flax. (After Flor 1947.)

Race	Ottawa	Bombay	F$_1$	Flax F$_2$			
	LLnn	*llNN*	*LlNn*	*L-N-*	*L-nn*	*llN-*	*llnn*
22 (*vLvL VNVN*)	S	R	R	R	S	R	S
24 (*VLVL vNvN*)	R	S	R	R	R	S	S
	Number of plants observed:			110	32	43	9
	Number of plants expected:			109	36	36	12

study. Virulence on each variety was determined by a recessive allele at a
single locus, and the two loci corresponding to each host resistance gene
were independent. Flor concluded that "the range of pathogenicity of a
physiologic race is determined by pathogenic factors specific for each
resistance factor possessed by the host" (Flor 1942).

Flor gave further examples in subsequent reports, but the most complete
information was presented in two papers. The first described the
inheritance of resistance to rust races 22 and 24 in the two flax varieties Ottawa 770B and Bombay (Flor 1947). The second described the inheritance
of virulence in these two races on the two varieties (Flor 1946). The two
sets of segregation data are given in Tables 4.1 and 4.2. *L* and *N* represent
dominant alleles for resistance in the host with recessive alleles *l* and *n*.
The symbols *vL* and *vN* represent recessive alleles for virulence in *M. lini*
with dominant alleles *VL* and *VN* for avirulence. In Ottawa flax, *L* determined resistance to race 24 but was ineffective against race 22, whereas
in Bombay flax, *N* determined resistance to race 22 but not to race 24. In
race 22 *vL* determined virulence on Ottawa, and in race 24 *vN* determined

TABLE 4.2
Inheritance of virulence on the flax varieties Ottawa 770B and Bombay in two races of
Melampsora lini. (After Flor 1946.)

Flax variety	Race 22	Race 24	F$_1$	F$_2$			
	vLvL VNVN	*VLVLvNvN*	*VLvLVNvN*	*VL-VN-*	*vLvL VN-*	*VL-vNvN*	*vLvLvNvN*
Ottawa (*LLnn*)	S	R	R	R	S	R	S
Bombay (*llNN*)	R	S	R	R	R	S	S
	Number of cultures observed:			78	27	23	5
	Number of cultures expected:			75	25	25	8

virulence on Bombay. In each case the F_2 ratios were close to those expected for two independent genes, showing that L and N and VL and VN were independent in each organism. The segregation ratios in the F_2 of the fungus are like those of a diploid organism because each culture is a dikaryon that originated from fertilization of a pycnium (see page 66).

Flor (1955, 1956a) summarized the evidence from flax rust in the gene-for-gene hypothesis. This states that during their evolution host and parasite develop complementary genic systems—that "for each gene conditioning rust reaction in the host there is a specific gene conditioning pathogenicity in the parasite" (Flor 1956a).

Several authors have pointed to the fact that with two alleles at a locus controlling resistance and two at a complementary locus controlling virulence, there are four possible interactions. Their tabulation is sometimes called a quadratic check. Only one of them—the combination of resistance and avirulence—leads to the expression of resistance and avirulence or, to combine the two viewpoints, a low infection type. (See (a) in Figure 4.1.) There is nothing remarkable about this. Apart from the fact that two organisms are involved, it is exactly comparable to the situation of two alleles at a locus determining auxotrophy or prototrophy in an organism which may encounter substrates with or without the growth requirement imposed by one of the alleles. (See (b) in Figure 4.1.) Growth fails in only one of the four combinations. The comparison of drug-resistant and sensitive strains on substrates with and without toxic concentrations of the drug also shows failure of growth in only one of the four combinations. (See (c) in Figure 4.1.) Although these examples could be models for the nature of specific resistance and how it is overcome, most interactions seem to be more complicated. In the host-parasite interaction, as in the other examples, no matter what other genes favor

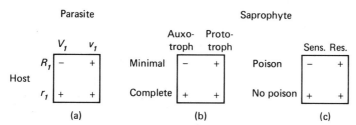

FIGURE 4.1

The nature of the substrate determines growth ($+$) or no growth ($-$) of (a) a parasite through interaction of host genes for resistance and parasite genes for virulence, and of saprophytes either by (b) the availability of a growth requirement, such as adenine, or (c) the presence of a toxic material, such as the arginine antimetabolite canavanine.

parasite growth and susceptibility, if any one does not, and so blocks growth, then resistance is expressed.

GENERAL VALIDITY OF THE CONCEPT

Since 1942 the gene-for-gene concept has been either demonstrated or suggested for a number of host-parasite systems. Their extent is shown in Table 4.3. An important feature of all these systems is that the host is an economic crop and that resistance is oligogenic. None are in any sense "natural." I shall return to this point later.

TABLE 4.3

Host-parasite systems in which a gene-for-gene relationship has been demonstrated (d) or suggested (s).

System	Reference
Rusts	
Linum–Melampsora lini (d)	Flor (1942)
Zea–Puccinia sorghi	Flangas & Dickson (1961) Hooker & Russell (1962)
Triticum–P. graminis tritici (d)	Green (1964, 1966) Kao & Knott (1969) Loegering & Powers (1962) Williams, Gough, & Rondon (1966) Luig & Watson (1961)
Triticum–P. striiformis (s)	Zadoks (1961) Lewellen et al. (1967) Line, Sharp, & Powelson (1970)
Triticum–P. recondita (d)	Samborski & Dyck (1968) Bartos, Dyck, & Samborski (1969)
Avena–P. g. avenae	Martens et al. (1970)
Helianthus–P. helianthi	Sackston (1962) Miah & Sackston (1970)
Coffea–Hemileia vastatrix (s)	Noronha-Wagner & Bettencourt (1967)
Smuts	
Triticum–Ustilago tritici	Oort (1963)
Avena–U. avenae	Holton & Halisky (1960)
Hordeum–U. hordei (d)	Sidhu & Person (1971, 1972)
Bunts	
Triticum–T. caries	Metzger & Trione (1962)
Triticum–Tilletia contraversa	Holton et al. (1968)
Mildews	
Hordeum–Erysiphe graminis hordei (d)	Moseman (1957, 1959)
Triticum–E. graminis tritici (d)	Powers & Sando (1957)

TABLE 4.3 *(continued)*

System	Reference
Other parasitic fungi	
Malus-Venturia inaequalis (d)	Boone & Keitt (1957), Day (1960)
Solanum-Phytophthora infestans (s)	Black et al. (1953) Toxopeus (1956)
Lycopersicon-Cladosporium fulvum (s)	Day (1956)
Solanum-Synchytrium endobioticum (s)	Howard (1968)
Nematodes	
Solanum-Heterodera rostochiensis (s)	Jones & Parrott (1965)
Insects	
Triticum-Mayetiola destructor (d)	Hatchett & Gallun (1970)
Bacteria	
Gossypium-Xanthomanas malvacearum (s)	Brinkerhoff (1970)
Leguminosae-Rhizobium (Symbiosis) (s)	Nutman (1969)
Viruses	
Lycopersicon-TMV (s)	Pelham (1966)
Lycopersicon-spotted wilt (s)	Day (1960)
Solanum-potato virus X (s)	Howard (1968)
Higher plant parasites	
Helianthus-Orobanche	Pustovojt (1965)

There are many more examples of parasites in which physiologic races have been defined in terms of oligogenic specific resistance. Although this in itself is circumstantial evidence of a gene-for-gene relationship, a formal proof still requires a demonstration comparable to that in flax rust. Two of the most recent proofs are interesting for the way that they overcame inherent technical problems in the genetical analyses.

The Barley-*Ustilago hordei* System

Plants of barley that are infected with covered smut (*Ustilago hordei*) produce masses of teliospores in place of grain. The spores are dispersed during threshing, carried on the surface of the seeds, and sown with them. In the soil the spores germinate at the same time as the host seeds, undergo meiosis, and give rise to compatible haploid products that fuse to form an obligately pathogenic dikaryon that systemically infects the host seedling. The dikaryon is never propagated by spores. The genetic variability inherent in the inoculum makes it necessary to isolate compatible haploid lines which can be separately cultured and paired to produce a genetically uniform inoculum when needed. Virulence is scored by estimating the percentage of smutted heads formed when inoculated seedlings reach flowering. Sidhu and Person (1971) showed that virulence on the barley varieties Hannchen and Vantage is determined by the gene *Uhv-1*, and that virulence on Excelsior is determined by *Uhv-2* (see page 78). These genes are independent and recessive, and for convenience we will write them as *v1* and *v2*, respectively, with dominant alleles *V1* and *V2* for avirulence.

The same authors (Sidhu and Person 1972) set out to test the premise that resistance in Hannchen and Vantage was controlled by a single dominant gene, that resistance in Excelsior was controlled by a second independent but also dominant gene, and that there is a specific gene-for-gene relationship between the host and parasite. They made the crosses Excelsior × Vantage and Excelsior × Hannchen, and raised F_1 and F_2 progenies. Scoring segregations in the F_2 for resistance to the two pathogen genotypes *v1v1 V2V2* and *V1V1 v2v2* introduced two problems. There was only one opportunity to inoculate a seedling, and even when seedlings were susceptible, about 1 in 5 would not form a smutted head. The second problem was that even if both pathogenic dikaryons were synthesized together from sporidia (for example, a mixture of *a1v1v2* + *a2v1V2* + *a2V1v2* where *a1* and *a2* are mating type alleles) and used as the

TABLE 4.4

Predicted ratios for F_3 barley lines inoculated with two dikaryons of *Ustilago hordei*. (After Sidhu & Person 1972.)

F_2 genotypes		Disease reactions in F_3 lines to: $v_1v_1\ V_2V_2$	$V_1V_1\ v_2v_2$		Summary of F_3 reactions
1/16 R_1R_1	R_2R_2	all R	all R	}	1/16 R R
2/16 R_1R_1	R_2r_2	3R:1S	all R	⎱	
1/16 R_1R_1	r_2r_2	all S	all R	⎰	3/16 S R
2/16 R_1r_1	R_2R_2	all R	3R:1S	⎤	
1/16 r_1r_1	R_2R_2	all R	all S	⎬	3/16 R S
4/16 R_1r_1	R_2r_2	3R:1S	3R:1S	⎦	
2/16 R_1r_1	r_2r_2	all S	3R:1S	⎤	
2/16 r_1r_1	R_2r_2	3R:1S	all S	⎬	9/16 S S
1/16 r_1r_1	r_2r_2	all S	all S	⎦	

R_1 and R_2 are hypothetical independent dominant genes for resistance from Hannchen and Vantage (R_1) and Excelsior (R_2).

inoculum, the F_2 ratio expected would be 9 resistant to 7 susceptible, provided there was no interaction in which a resistant reaction would prevent the development of a virulent component. A 9:3:3:1 ratio is a far more sensitive test of the hypothesis. Both problems were overcome by selfing F_2 plants to give separate F_3 lines. The seed of each F_3 line was divided into two lots, one inoculated with $v1v1\,V2V2$, the other with $V1V1v2v2$. Since the F_3 lines were themselves segregating, they were scored as susceptible if any smutted heads were found, and only scored as resistant if none was found. Although the F_2 segregation expected is 9:3:3:1 for plants resistant to both, to one, to the other, or to neither pathogen line, the corresponding F_3 ratios will be 1:3:3:9, using the criteria just explained. This relationship is shown in Table 4.4. The results obtained are given in Table 4.5 and are very close to those expected from the hypothesis. The genes for resistance are dominant and independent and interact with genes in the pathogen as the gene-for-gene hypothesis predicts.

TABLE 4.5

F_3 ratios from two barley crosses inoculated with two dikaryons of *Ustilago hordei*. (Data from Sidhu & Person 1972.)

Expected reaction of F_3 lines with $v_1v_1\ V_2V_2$	$V_1V_1\ v_2v_2$	Expected frequency	Excelsior x Hannchen Observed	Expected	Excelsior x Vantage Observed	Expected
R	R	1/16	13	9	15	9
R	S	3/16	30	27	32	28
S	R	3/16	28	27	31	28
S	S	9/16	72	80	70	83
			143	143	148	148

The Wheat-Hessian Fly System

The control of insect pests of crop plants by resistance has been practiced for over a century. One of the most successful and earliest examples is the control of the grape louse *Phylloxera vitifoliae* in Europe with resistant vines from North America (Riley 1872). Painter (1951, 1966) has reviewed the literature on insect resistance up to 1951. By 1966 he had noted six examples in which biotypes or races able to feed on resistant varieties had developed. These are equivalent to physiologic races in plant pathogenic fungi. It would have been surprising if insects had not developed such forms. Under favorable conditions they can produce very large numbers of individuals, and are no less versatile genetically than the fungi. The common occurrence of insecticide resistance shows how well equipped they are to meet man's efforts to control them (see page 156). One of the best examples of the development of races in an insect pest is the Hessian fly *Mayetiola destructor*. Recent work by Hatchett and Gallun (1970) has shown that a gene-for-gene relationship exists in this host-pest interaction. Races were first found by Painter in the early 1930's, who noted that Hessian fly larvae from western Kansas did not flourish on certain soft wheats, whereas larvae from eastern Kansas did. Active wheat breeding for Hessian fly resistance followed, and by 1959 five dominant genes for resistance had been isolated, and with other unidentified genes, released in wheat varieties. By 1969 a total of 26 resistant wheats had been released and most of them used in 34 states (Gallun and Reitz 1971). For example, 10 varieties, representing two of the resistance genes, together made up 93 percent of the wheat acreage in Indiana. The consequences are of considerable interest since they show that what is now commonplace with pathogenic fungi also occurs with plant-eating insects.

Before discussing the genetic studies, we must consider some features of the biology of the Hessian fly. Painter (1951) gave a full account of the early work, and Gallun et al. (1961) the more recent work. Female Hessian flies lay their eggs on wheat seedlings (Figure 4.2:B). The eggs are deposited in rows on the leaf surface in the longitudinal depressions between adjacent vascular bundles. The newly hatched larvae crawl down inside the base of the leaf sheath. Here they are believed to feed by secreting enzymes and ingesting materials which diffuse through the host cell walls. If an infested seedling is susceptible, its leaves become stunted and turn dark green, and new leaves may fail to form. Resistant seedlings show some leaf-stunting at first, but they generally recover, and in any case, remain light green like uninfested plants. Larvae feeding on resistant plants generally die, but some may survive (remaining small in size) and fail to stunt the seedlings. At this point we can borrow the terms virulent and avirulent to distinguish biotypes that are able from those that are unable to feed on a resistant host.

Populations from single-pair matings between flies can be maintained in the laboratory. Progenies are tested for race phenotype by scoring plant reaction to larval feeding and the ability of the insect to survive. A pair of flies are released in a cheesecloth cage placed over wheat seedlings at the two-leaf stage. Seedlings of several genotypes may be grown in a single pot. The female lays eggs at random, showing no host preference. The larvae are scored as dead or alive 15 days after egg laying has ceased, and the lengths of those that are alive are measured. The seedling reactions provide an indication of the progeny phenotype (Figure 4.2: C and D). Resistant seedlings infested with segregating progenies appear susceptible even if only one virulent larva is present on a plant. Thus, resistant reac-

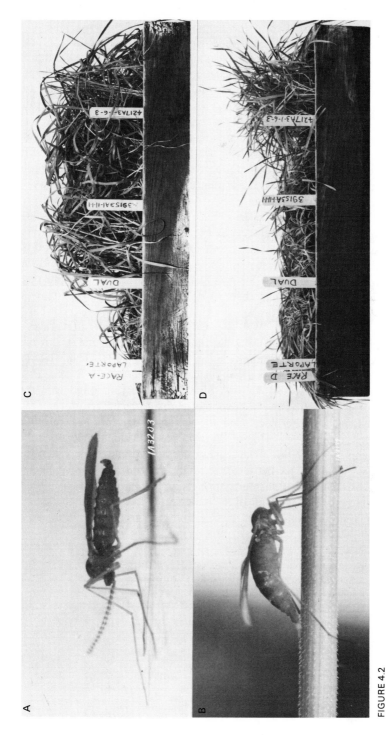

FIGURE 4.2

Hessian fly (*Mayetiola destructor*) A—male, B—female laying eggs; C and D differential reaction of seedlings of 4 wheat varieties to caged populations of races A and D. A and B from Kansas Agricultural Experiment Station; C and D from Gallun et al. (1961).

tions do not interfere with the development of virulent larvae, and in fact the reverse may occur. The close proximity of virulent larvae appears to promote the survival of avirulent siblings. Hatchett (personal communication) has suggested that virulent larvae inactivate, or neutralize, an antibiotic factor thought to be responsible for resistance. In segregating progenies, larval size distinguishes virulent (large) larvae from avirulent (small) larvae.

Hessian fly genetics is complicated by the fact that when male flies form sperm, the paternal haploid genome, or chromosome set, (n = 8) is eliminated and only the maternal chromosomes are transmitted (Gallun and Hatchett 1969). As a consequence F_2 and backcross progeny genotypes from crosses between adults of different races vary according to the direction of the original cross between the parents. F_1 flies used as males breed as though homozygous because they transmit only maternal chromosomes. The F_1 females breed as heterozygotes, showing normal transmission of both genomes subject to reassortment and crossing-over at meiosis. A similar transmission was recently reported for the male of the lac insert *Kerria lacca* (Chauhan 1970). All the progeny of a single-pair mating in Hessian fly are of the same sex.

Hatchett and Gallun (1970) studied crosses between three races of Hessian fly, distinguished by the reactions shown in Table 4.6. Monon wheat carries the single dominant gene for resistance H_3 (Caldwell et al. 1946), whereas the resistance of Seneca is undetermined. Race GP, the great plains race, is avirulent on all wheats carrying oligogenic race-specific resistance. Table 4.7 shows the segregations observed when races GP and E were intercrossed, and F_1, F_2, and backcross progenies were tested on seedlings of Monon. Avirulence is evidently dominant to virulence in the F_1 hybrid. The segregating F_2 and backcross generations show that the difference between races GP and E is due to a single gene. Those crosses in which no segregation occurred were the result of elimination of the paternal genome of the male parent as explained above.

From a similar analysis of F_1, F_2, and backcross data from crosses

TABLE 4.6
Differential reactions of 3 races of Hessian fly. S = high, or susceptible; R = low, or resistant. (From Hatchett 1969.)

| Wheat variety | Genotype | Hessian fly races | | |
		GP	A	E
Turkey	h	S	S	S
Seneca	?	R	S	R
Monon	H_3	R	R	S

TABLE 4.7
Phenotypes of larvae from progenies of reciprocal crosses between Hessian fly races GP (Great Plains) and E determined on Monon wheat, carrying the incompletely dominant gene H_3 for resistance. (Hatchett and Gallun 1970.)

	Cross ($♀ \times ♂$)	No. of progenies	No. of larvae	avirulent (GP) %	virulent (E) %	ratio
				Phenotypes of larvae		
F_1	GP $\times$ E	39	3381	100	0	—
	E $\times$ GP	36	2195	100	0	—
F_2	(GP $\times$ E) $\times$ (GP $\times$ E)	35	2271	100	0	1:0
	(GP $\times$ E) $\times$ (E $\times$ GP)	13	1351	50.8	49.2	1:1
	(E $\times$ GP) $\times$ (E $\times$ GP)	27	1656	52.2	47.8	1:1
	(E $\times$ GP) $\times$ (GP $\times$ E)	26	3066	100	0	1:0
BC	(GP $\times$ E) $\times$ E	13	1329	40.2	59.8	1:1
	E $\times$ (GP $\times$ E)	24	1132	100	0	1:0
	(E $\times$ GP) $\times$ E	11	866	46.9	53.1	1:1
	E $\times$ (E $\times$ GP)	28	2065	0	100	0:1

between races GP and A, it was concluded that virulence on Seneca was also determined by a single recessive gene. Crosses between races E and A gave the results shown in Table 4.8. Since each larva can only be tested once, separate progenies were established on the differential varieties Seneca and Monon.

The segregations show that two independent genes condition reaction type on Monon and Seneca. F_1 hybrids between A and E are avirulent on both varieties and thus have the phenotype of race GP. We may note that because paternal chromosomes are not transmitted by the male, only the

TABLE 4.8
Phenotypes of larvae from progenies of reciprocal crosses between Hessian fly races A and E determined from tests on two wheat varieties. (Hatchett and Gallun 1970.)

	Cross ($♀ \times ♂$)		larvae on Seneca			larvae on Monon	
		No.	% virulent (A)	% avirulent (E)	No.	% avirulent (A)	% virulent (E)
	A $\times$ A	532	100.0	0	525	100.0	0
	E $\times$ E	309	0	100.0	739	0	100.0
F_1	A $\times$ E	1042	.4	99.6	1151	97.6	2.4
	E $\times$ A	1176	.8	99.2	1169	94.3	5.7
F_2	(A $\times$ E) $\times$ (A $\times$ E)	270	49.3	50.7	248	98.0	2.0
	(E $\times$ A) $\times$ (E $\times$ A)	384	2.2	97.8	1061	28.1	67.4

phenotypes of the original maternal parent and GP are recovered in each F_2 progeny, and never both parental phenotypes or the double recombinant. In fact, the double recombinant will only appear at a frequency of 3/64 in the F_3, whereas the original male parent phenotype appears at a frequency of 2/64 or 1/32.

The data of table 4.8 are most readily explained if the wheat variety Seneca carries a single gene for resistance to Hessian fly corresponding to the gene for avirulence in race E. Hatchett and Gallun (1970) noted that the genetic basis of resistance in Seneca was unknown. Recent genetic studies of crosses between Monon and Seneca evaluated for resistance to races C and E suggest that Seneca carries two genes for resistance to race E (Gallun and Patterson: personal communication). These are provisionally designated H_1 and H_2 on the basis of their similarity to the genes H_1 and H_2 in Dawson wheat (Noble and Suneson 1943). There are two simple explanations for the discrepancy. One is that the two genes in Seneca may be duplicate genes or, in other words, H_1 and H_2 have identical specificities and belong to different genomes. The other is that whereas the race E culture used in scoring the Monon x Seneca progenies carried two avirulence genes ($A_1A_1 A_2A_2$), the race E culture used in the cross with race A only carried one ($A_1A_1 a_2a_2$ or $a_1a_1 A_2A_2$). Either explanation is consistent with the gene-for-gene hypothesis. A third possibility, that the avirulence gene of race E has a dual specificity (for H_1 or H_2), would be the first example of its kind. Further tests will no doubt soon resolve the matter.

At the time of this writing, all the 8 races which can be defined by the 3 different resistant varieties Seneca, Monon, and Knox 62 are known (Table 4.9). This situation closely parallels that described for fungal

TABLE 4.9
Races of Hessian fly known in 1972. Variety Seneca provisionally assigned genes H_1 and H_2 described from Dawson. (Based on Hatchett and Gallun 1970, and Gallun and Reitz 1971.)

Wheat variety	Resistance gene	GP	A	E	F*	B	G*	C	D
Turkey	h	S	S	S	S	S	S	S	S
Seneca	H_1H_2	R	S	R	R	S	R	S	S
Monon	H_3	R	R	S	R	S	S	R	S
Knox 62	H_6	R	R	R	S	R	S	S	S
Ribeiro	H_5	R	R	R	R	R	R	R	R

*Races F and G only known from laboratory populations.

pathogens. Almost certainly many of the conclusions drawn from plant pathology concerning the future of genes for race-specific resistance and the part they play in epidemiology will apply to this and other insect pests.

HEURISTIC VALUE OF THE GENE-FOR-GENE CONCEPT

Host Genotypes

The alleles for avirulence and virulence at a locus in the pathogen are distinguished in terms of a host gene for resistance. This means that if a pathogen culture carrying a single allele for avirulence on a given host variety is also avirulent on other varieties, by virtue of the same allele, then all those varieties will have a common gene for resistance. The array of reactions of a group of different pure lines or clones of host material to pathogen segregants can thus be used to predict putative host genotypes for resistance.

Table 4.10 shows some predictions for the *Malus-Venturia inaequalis* system that are based on the type of interactions shown for the variety Haralson in Figure 3.5. One parent culture of *Venturia* was virulent on the host variety being tested and the other was avirulent. The 8 haploid cultures from a single ascus show the segregation of these phenotypes. Several of the predictions shown in the table were confirmed. For example, the apple varieties "McIntosh" and "Hyslop" were found to carry resistance genes corresponding to avirulence alleles at *p-1* and *p-5* (Boone 1971). Bagga and Boone (1968b) showed that resistance to a particular culture in "Almata," "Prairie Crab," "Oekonomierat-Echtermeyer," and "Red Silver" was in each case determined by one gene. The test cross progenies for "Hopa," "Jay Darling," and "Morden 451" were expected to show segregation of two genes for resistance to *Venturia* culture 365–4. However, in each case the second resistance gene had been "identified" with another avirulent culture of *Venturia*. Since culture 365–4 carried avirulence alleles at loci *p-14* and *p-16* but virulence alleles at *p-15* and *p-17*, segregation of R15 and R17 was not detected.

Where a large amount of information is available, such as the reactions of up to 60 or more host lines to a comparably large number of pathogen cultures, computer methods are useful in establishing genetic homologies. By this means Loegering et al. (1971) classified 60 wheat varieties into several groups with common resistance genes, making use of their reactions to 9 different races of *Puccinia recondita*. Dinoor (personal communication) has also developed a computer program to establish putative resistance genotypes in breeding lines and wild collections of oats inoculated with races of *Puccinia coronata*.

TABLE 4.10

Resistant reactions (x) controlled by alleles for avirulence at 19 loci in *Venturia inaequalis* on 20 apple varieties.

Apple variety (reference)	Reactions of Venturia stocks carrying alleles for avirulence at loci							Putative host genotype
	p-1	*p-2*	*p-3*	*p-4*	*p-5*	*p-6*	*p-7*	
(Boone & Keitt 1957)								
McIntosh	x							*R1*
Yellow Transparent	x		x	x				*R1 R3 R4*
Haralson		x		x				*R2 R4*
Red Astrachan			x					*R3*
Hyslop				x	x			*R4 R5*
Grimes Golden						x		*R6*
Prairie Spy		x					x	*R2 R7*

	p-8*	*p-9*	*p-10*	*p-11**	*p-12**	*p-13**	
(Williams & Shay 1957)							
Dolgo	x						*R8*
Russian		x					*R9*
Geneva			x	x			*R10 R11*
Malus sikkimensis					x	x	*R12 R13*

	p-14	*p-15*	*p-16*	*p-17*	*p-18*	*p-19*	
(Bagga & Boone 1968a)							
Almata	x						*R14*
Hopa	x	x					*R14 R15*
Wild Sweet Crab		x					*R15*
Prairie Crab			x				*R16*
Jay Darling			x	x			*R16 R17*
Morden 451			x	x			*R16 R17*
Oekonomierat Echtermeyer					x		*R18*
Red Silver						x	*R19*
Scugog						x	*R19*

*Loci whose alleles determine somewhat less extreme phenotypic differences than the others.

Physiologic Races

The gene-for-gene hypothesis also predicts the number of physiologic races a set of resistant varieties will differentiate. With n genes for resistance, each of which may have two phenotypes (resistant or susceptible), there are 2^n different races. Thus 19 resistance genes in apples would define 2^{19}, or 524,288, races of *Venturia*. In fact, the host varieties listed in Table 4.10

will not distinguish that many races because 9 of them carry two or more resistance genes. Monogenic differentials are required to identify all the races possible (Person 1959).

The identification and naming of all the races of *Venturia* would be a rather futile exercise, but there are some races that are of practical interest. Williams and Kuć (1969) noted the occurrence of races virulent on clones of certain apple seedling selections with monogenic resistance that were released in the 1950's and '60's, with high hopes of controlling apple scab. It seems that the original resistance of the source material owed much to minor genes that were not transferred in the breeding programs.

RACE NAMES

Arbitrary Numbers

Naming races by arbitrary designations of numbers of letters was a good enough method when there were few resistance genes in use and not many races. As new resistant cultivars were introduced, new races were identified and supplemental differentials had to be added. Old race numbers were modified by the addition of letters or other numbers, and tables of reactions rapidly became incomplete and out-of-date. The patching and modification to bring old schemes up-to-date is evident in race designations such as 15B-1LX (Can.) or 13 AFH, two recent examples from *P. graminis tritici* and *P. graminis avenae,* respectively. Several alternative systems have been proposed. Some depend on using single-gene differentials recognizing only two phenotypes, resistant or susceptible. Others use numbered host varieties that need not be fully characterized genetically. Still others can take account of detailed reaction phenotypes with the aid of data-processing machines.

Race Formulae

The first scheme that did not use arbitrarily assigned numbers was proposed by Black et al. (1953) for international designations for races of *Phytophthora infestans* on potatoes. It was very successful. The genes for resistance were numbered *R1, R2, R3,* and so forth, to *R11* (see page 49), and the races were numbered to indicate their virulence towards them. Race 0 was avirulent on all resistant varieties, whereas race 1,2,3 was virulent on any combination of the genes *R1, R2,* and *R3,* and so on. A similar scheme was suggested for the tomato-leaf mold disease *Cladosporium fulvum* (Day 1956). The conceptual simplification was useful in breeding for resistance so long as the numbers of genes remained small so that the race designations did not become unwieldy.

Similar ideas have been put forward for races of cereal rusts to augment the standard race designations (Johnson et al. 1967). The first such proposal was by Simons and Murphy (1955), who established descriptive formulae for races of crown rust on 13 differential oat varieties assigned the numbers 1 to 13. Thus race 29 had the formula 1,2,3,4,5,6,7,8,9,10,11, 12,13, showing that it was avirulent on all 13 differentials. Race 101 with the formula 13 was avirulent only on variety number 13 (Glabrota). The oat crown rust race formulae thus emphasize avirulence, whereas the potato late blight race names emphasize virulence.

Watson and Luig (1961, 1963) adapted the *P. infestans* scheme to deal with the problem of supplemental differentials to identify races of *Puccinia recondita* and *P. graminis tritici*. Standard race numbers were determined for both rusts by using the established international differentials, but for each rust a set of locally important supplementary differentials were also used. For example, the supplementary differentials for *P. graminis tritici* were the wheat varieties McMurachy-Sr6 (1), Yalta-Sr11 (2), W2402-Sr9 (3), C.I.12632 (4), Renown (5), and W2401-Sr8 (6). These were assigned numbers, shown in parentheses. The overall race designation was given by the standard race number followed by the letters Anz to identify the geographical area of Australia and New Zealand, followed by the numbers of the supplementary differentials that were susceptible. Thus races 126-Anz-6, 126-Anz-1, 6, and 126-Anz-2,6 are not discriminated by the standard differentials; they are all race 126 by that test, but they are separated by their virulence on one, two, or three of the supplementary differentials.

The next step was the use by Green (1963, 1965) of "virulence formulae" for races of wheat stem rust. For example, the formula 6,7,10/5,8,9a,9b,11 indicates a group of races that are avirulent on wheats carrying any of the rust resistance genes Sr6, Sr7, or Sr10, but are virulent on wheats that are only protected by any one or more of the remaining resistance genes. The symbol / separates the two classes of genes: effective to the left and ineffective to the right. The formula can be referred to by its formula number (C15), which then replaces the original race name. This method, or a variant of it, is currently used for races of stem rust (Green 1963, 1971a), leaf rust *Puccinia recondita* (Samborski 1969), and yellow rust *P. striiformis* (Line et al. 1970).

The most comprehensive scheme proposed so far combines the elements of the other schemes together to designate races of *P. recondita* on the basis of sequentially numbered wheat lines (Loegering and Browder, 1971). Each investigator determines virulence formulae using these numbers rather than numbered resistance genes. This has the advantage that differentials can be used before the genetic basis of their resistance has been established. The formulae are assigned temporary formula codes designating the year and the laboratory, or state, where the determination was made.

Binary Notation

Habgood (1970) recently proposed a novel and ingenious system of race naming. A unique decanary, or base 10, number is generated for each race from a binary number. The latter is derived from the array of "0" or "1" alternatives, representing resistant or susceptible reactions of a set of differentials in a fixed linear order. For example, a pathogen culture inoculated to a series of eight differentials (A to H) may be virulent on B, D, E, and F:

Differential host	H	G	F	E	D	C	B	A
Reaction	R	R	S	S	S	R	S	R
	0	0	1	1	1	0	1	0
Binary value	(2^7)	(2^6)	2^5	2^4	2^3	(2^2)	2^1	(2^0)
Decanary value			32	16	8		2	
Designation of culture = 32 + 16 + 8 + 2 = race 58								

When new differentials are needed, they are simply added to the left of the fixed order. As Habgood points out, there are additional benefits; for example, only odd numbered races are virulent on variety A, or only races with numbers of 64 or greater are virulent on variety G. The chief drawback, the difficulty in going from binary to decanary or back, would no doubt disappear with practice. Another drawback is having to retain the early differentials in the series when they have outlived their usefulness.

Habgood's proposal has been applied to naming races of *Puccinia striiformis* (Johnson et al. 1972). The nomenclature is also based on numbers assigned to the differentials themselves rather than the resistance genes they carry. Two different series of differentials were proposed: a set of 7 international differentials, and an entirely different set of 8 European differentials. A third set could be used to take care of still more localized specialization. It was also suggested that races should be designated by a world number followed by a local number with a letter prefix to indicate the region. Using several sets of differentials neatly overcomes the problem of cumbersome race designations.

The earliest race-numbering schemes allowed, in theory, the flexibility of recognizing and using grades between the extremes of resistance and susceptibility. In practice, it was simpler and tidier to say, for example, that rust reactions from 0 to 2 were resistant and from 3 to 4 were susceptible. The heterogeneous class (5 or X) was considered resistant. It was certainly easier to profit from the gene-for-gene concept by regarding each resistance gene as either effective or ineffective and ignoring intermediates. In any case varieties whose reactions were not clear-cut were most unlikely to be included in sets of differentials.

The next developments will no doubt be in the direction of regarding race designations as largely unnecessary. In the race surveys that are carried out each year, the most important information is the frequency of certain critical virulence genes in the pathogen population. These are the genes that render ineffective the major resistance genes currently deployed in the crop. If combinations of two or more resistance genes are deployed, then the corresponding combinations of virulence genes will also be critical. Virulence formulae provide this information and lend themselves to computing gene frequencies. The methods required for seed sowing, inoculation, and record keeping on a scale large enough to be effective were developed for *Puccinia recondita* by Browder (1971 a,b,c).

AN OVERSIMPLIFICATION

Although the gene-for-gene hypothesis is a useful conceptual tool for many plant-parasite systems, its application is not universal. Where it has been exclusively applied to specific resistance to parasites, there is, in theory, no reason to suppose that it does not also apply to general resistance. Although there is still little experimental evidence for polygenic control of virulence among parasites, it would be very surprising not to find it (see pages 187–189).

One somewhat ironical view of the gene-for-gene concept is that it is an artifact of breeding for resistance. Plant breeders stripped oligogenic resistance of associated protective polygenic effects and exposed the genes one at a time in agriculture on a tremendous scale, forcing the parasite to respond in kind. Johnson (1961) called this "man-guided" evolution. It is perhaps of some significance that the gene-for-gene concept has not been demonstrated in nature, although it probably will be eventually. A likely example is the *Melandrium-Ustilago violacea* system where Goldschmidt's (1928) pioneer studies showed that virulence on different host species among 6 races of the smut fungus was under simple genetic control.

The artificiality of crop monoculture and the problems posed by genetic uniformity are widely appreciated, and there are now some programs to alleviate, or remedy, the consequences that are discussed in the last chapter.

GENE FUNCTION IN
HOST-PARASITE INTERACTION

Our discussion so far has dealt with the inheritance of resistance in host plants and virulence in parasites and how the interaction of the two genetic systems can often be described, at a simple level, by the gene-for-gene hypothesis. We must now consider what is known of the nature of resistance and virulence, and what this can tell us about the functions of the genes involved. We will first briefly examine the biochemical basis of disease resistance, noting the general patterns that have emerged in recent years, and then discuss some theories on the nature of resistance and virulence.

NATURE OF RESISTANCE TO PARASITES

Most plants have ways of either avoiding or lessening the impact of some of their potential pathogens. Early maturity, completion of growth and the reproductive cycle before an annual wet season, or tolerance of conditions which adversely affect parasite development, such as extremes of temperature or humidity, are examples of avoidance mechanisms. They are genetically determined and may be very useful in breeding for disease control, but will not be mentioned further in this chapter.

A host that is unable to avoid direct confrontation with a parasite but is nevertheless able to restrict its development is resistant (Robinson 1969). There are several kinds of resistance: it may be physical or chemical, it may depend on the lack of a nutrient or substance the parasite needs for development, or it may be due to a toxic or repellant substance that is either preformed or formed only in response to infection.

Physical Resistance

Some kinds of physical resistance are very striking. The vine *Passiflora adenopoda* has foliage covered with small hooklike hairs that effectively hold and trap the larvae of butterflies of the genus *Heliconius* by hooking into their prolegs (Gilbert 1971). The larvae die of starvation and also by loss of haemolymph through the numerous puncture wounds in their integuments. This seems a very successful method of controlling a parasite and is only likely to be circumvented by larvae with harder or tougher skins.

Adhesives can also immobilize insect predators and are formed by several species of *Solanum* (Gibson 1971). Resistance to aphid infestation in *S. polyadenium* and two other species is due, at least in part, to an exudate discharged by abundant glandular hairs on the epidermis. When ruptured, the hairs release a clear fluid which hardens on exposure to the air, forming a black insoluble material. Aphids were placed on plants of these potato species, and about 30 percent became stuck within 24 hours. Similar glandular hairs are found on tomatoes and other Solanaceae, and may also provide aphid resistance.

Physical resistance to fungal infection is sometimes due to increased cuticle thickness or to the distribution of thickened mechanical tissues like sclerenchyma. Wood (1967) has described other examples.

Resistance to Phytotoxins

Several important pathogenic fungi produce host-specific toxins whose effects determine the nature of host-parasite interaction. The effects include an increase in the loss of electrolytes that appears to result from changes in membrane permeability. For example, membrane-bound naked protoplasts prepared from coleoptiles of susceptible oat seedlings lyse or burst in solutions of the specific toxin of *Helminthosporium victoriae* (HV-toxin), whereas protoplasts from resistant seedlings are unaffected (Samaddar and Scheffer 1968). The genetic basis of resistance to two toxin-producing fungi (*H. victoriae* and *Periconia circinata*) was discussed on pages 17–20, and the genetic control of toxin production in *H. victoriae* and several other species was dealt with on pages 53–55.

It appears that resistance to host-specific toxins does not depend on induced biochemical activity on the part of the host. Much of the evidence for this view was presented in a recent review by Scheffer and Yoder (1972). Resistant and susceptible oat tissues both inactivate HV-toxin at comparable rates so that resistance is not due to toxin inactivation. Dormant seeds of a susceptible line briefly exposed to toxin solutions, redried, and stored until no more toxin was detectable failed to germinate. Seeds of a resistant line similarly treated germinated normally. In dormant half-seeds, with no embryos, it was also possible to show that aleurone tissue treated in a similar fashion could synthesize α-amylase when activated with gibberellic acid, provided the seed was from a resistant line. Rates of respiration and protein synthesis are extremely low in dormant seeds, and yet resistance was expressed. Aleurone from seeds of a susceptible line failed to synthesize α-amylase after toxin treatment.

It is only fair to point out that the nature of HV-toxin action is a subject of much controversy. Luke and Gracen (1972) have written a lucid and comprehensive account of the several hypotheses advanced to account for resistance and susceptibility. At the present time most workers agree that susceptible cells possess a toxin receptor or sensitive site that may be located on the plasma membrane. Susceptible oat cells appear to be damaged by concentrations of HV-toxin below 100 molecules per cell. They can be protected by certain agents which may induce changes in toxin receptor sites. These agents include uranyl salts (known to bind to membranes) and sulfhydryl-binding compounds. The remaining problems are to establish where the toxin receptor sites are and to see whether resistant tissue has any receptor sites, and if so, whether it has in addition the capacity to inactivate toxin or rapidly repair the damage it causes (Luke and Gracen 1972).

The nature of resistance to eyespot disease (*H. sacchari*) in sugarcane was recently established by Strobel (1973). *H. sacchari* produces a host-specific toxin identified as 2-hydroxy-cyclopropyl-α-D-galactopyranoside (helminthosporoside) that binds to a membrane protein of susceptible clones. Membranes prepared from resistant clones do not bind the toxin. The two kinds of membrane protein were prepared from a susceptible and a resistant clone and shown to vary with respect to four amino acid residues. The structural alteration in the membrane binding protein of the resistant clone is thus directly associated with its resistance to eyespot disease.

Scheffer and Yoder (1972) list 8 examples of fungal pathogens, known in 1970, whose toxins show the same host specificity as the pathogens themselves. Toxin mediated host-parasite interaction is clearly more common in plant diseases than was formerly thought.

So far, interactions based primarily on phytotoxins seem simple compared with other specific interactions. HV-toxin is only effective on oats carrying *Pc-2*. No strains of *Helminthosporium victoriae* are yet known that produce HV-toxins capable of damaging other oat lines. This may mean that toxin molecules with new specificities that would enlarge a pathogen's host range evolve much more slowly compared to the fairly common mutation to virulence considered later in this chapter.

For the rest of this chapter I discuss specific resistance. As its name implies, it can be overcome by the parasite. A central question in the study of host-parasite relations at the present time is why specific resistance is effective against some but not other forms of a parasite—in other words, the nature of specificity.

Defense Reactions

Most living plant tissues have a common response to physical injury. The damaged cells and some adjacent cells die, and in the process produce a variety of compounds, some of which are oxidized polyphenols, which are inhibitory to the growth of microorganisms. Damaged tissue becomes discolored, turning dark brown to black. Soon after injury, the rates of DNA, RNA, and protein synthesis increase in comparison with uninjured tissue. Respiration also increases, as does the activity of a variety of enzymes. The end point is the conversion of damaged tissue that would otherwise be a substrate for growth of saprophytic microorganisms into a toxic and unsuitable substrate, clearly delimited from the surrounding healthy tissue. The wounded area eventually dries, becoming necrotic, and in a leaf may even fall away leaving a hole. Many different physical or chemical insults can produce this wound response. The changes that occur in damaged tissue are like those that occur when tissues age and die during senescence. The chief difference is that the changes resulting from injury are more rapid and are localized. (See also pages 135–138.)

Many would-be parasites induce similar defense reactions, and as a result, are denied further access to host tissue because they are either killed or prevented from growing. The interaction was called hypersensitivity because it resulted in the death of one or more host cells at the site of penetration. Hypersensitive resistance was at first thought to be of value only against obligate parasites since they could not grow on dead host tissue. In fact, the accumulation of toxic materials in the reacting tissue, and not the death of host cells, is the critical factor.

Phytoalexins

A great deal of effort has been devoted to the identification and characterization of the compounds produced by plants in response to infection or infestation. During the last 30 years many papers have described the structure, synthesis, and activity of such compounds now called phytoalexins. The term was introduced by Müller and Börger (1940) to describe the substances that inhibit fungus development formed when living plant tissues are invaded by a fungal parasite. It has since been redefined and broadened to include substances formed by host tissue ". . . in response to injury, physiological stimuli, infectious agents or their products . . . [that] accumulate . . . to levels which inhibit the growth of microorganisms" (Kuć 1972). I propose also to include substances that are formed in response to infestation that are toxic to, or which repel, insect and nematode parasites.

Most of our detailed knowledge of phytoalexins comes from the study of rather few host tissues. These include cut surfaces of potato tubers and sweet potato roots, inner tissues of pea, green bean, and broad bean pods, carrot roots, soybean seedling hypocotyls, and sometimes the leaves of these and other host plants. The methods used generally involve exposure of tissue to a variety of agents including spores, mycelia, culture filtrates or extracts of pathogens and nonpathogens, and more recently, drugs, antimetabolites, heavy metal ions, and other potential inducers.

Controls to distinguish between materials formed in response to injury in tissue preparation and those that appear after treatment are generally included but are not always adequate. The materials formed in the tissue are detected by a variety of methods including gas chromatography. Unknowns are isolated, purified, and their structures determined by standard procedures. Their antibiotic activities are measured by bioassay.

The taxonomic distribution of the characterized phytoalexins was recently reviewed by Ingham (1972). The following account attempts only to relate phytoalexins to what is known of the genetic control of host-parasite interaction.

POTATO TISSUE

Müller and Börger (1940) and Müller (1953) were the first to show that potato tubers carrying the gene *R1* for late blight resistance responded when inoculated with an avirulent race (race 0) of *Phytophthora infestans* by producing phytoalexins that inhibited development of a virulent race (race 1) applied a short time later. A number of other examples of host-parasite systems in which preinoculation with an avirulent strain protects against subsequent infection with a virulent strain were recently reviewed by Matta (1971). Tomiyama et al. (1968) later showed that one of the principal phytoalexins formed by *R1* tubers was a bicyclic sesquiterpene alcohol given the trivial name of rishitin (Figure 5.1). Tissue inoculated with the virulent race 1 accumulated only 0.44 μg of rishitin per g of tissue (fresh weight), whereas tissue inoculated with race 0 formed 120 μg/g, and uninoculated sliced tissue formed only trace amounts of rishitin. Sato et al. (1968) showed that rishitin not only accumulated when tubers of potato varieties carrying the genes *R2, R3,* and *R4* were inoculated with race 0, but also when tuber tissue of a susceptible variety was treated with a cell-free extract of the fungus. Further work showed that boiled aqueous extracts of three different races of *P. infestans* caused accumulation of rishitin and a related compound, called phytuberin, in tuber slices of both resistant and susceptible varieties (Varns, Currier, and Kuć 1971). Two nonpathogens of potato (*Ceratocystis fimbriata* and *Helminthosporium carbonum*) also stimulated accumulation of these two compounds in both kinds of tuber tissue.

The phytoalexins rishitin and phytuberin are only sparingly soluble in water and are normally extracted using organic solvents. It is very difficult

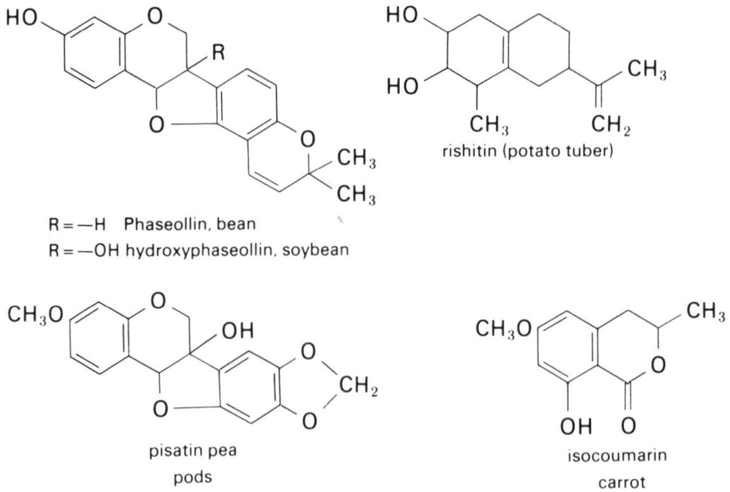

R = —H   Phaseollin, bean
R = —OH hydroxyphaseollin, soybean

rishitin (potato tuber)

pisatin pea
pods

isocoumarin
carrot

FIGURE 5.1
Some phytoalexins and their sources.

to establish the concentration of phytoalexins in tissue at the sites where they act, and Daly (1972) has questioned whether such materials do in fact accumulate to levels that are inhibitory.

Evidently, the *R* genes transferred to *Solanum tuberosum* from *S. demissum* are not required for the formation of rishitin and phytuberin in tuber tissue but rather determine the ability of the host tissue to respond to avirulent races of *P. infestans.* An avirulent race could produce a substance that interacts with an *R* gene product to initiate phytoalexin accumulation. If this is a characteristic of all specific resistant reactions, the number of such mechanisms for initiating phytoalexin synthesis must at least equal the number of *R* genes known in the potato (11) and may well be even greater.

There is some evidence that virulent races differ from avirulent races in other ways than by the absence of an inducer. Tomiyama (1966) found that when potato petiole tissue was inoculated with a virulent, or compatible, race of *P. infestans,* neither host nor parasite showed signs of death or damage for a period of 2 days. After this time, degenerative changes began in the host tissue. In contrast, inoculation with an avirulent, or incompatible, race was followed by death of some infected individual cells within 10 minutes of penetration. By 40 to 60 minutes, most penetrated cells were dead (Tomiyama 1967). When petiole tissue that had been inoculated with a compatible race was inoculated 15 to 20 hours later with an incompatible race, no host cell death occurred within 4 hours. This was the length of time in which the two inocula could still be distinguished by their relative amounts of growth. Evidently the virulent race in some way suppressed the ability of the host cells to show a hypersensitive response.

Varns and Kuć (1971) showed that resistant tuber tissue inoculated with a virulent race does not form rishitin and phytuberin in response to inoculation with an avirulent race applied 12 or more hours later. Examples of mixed lesions of virulent and avirulent races in other host-parasite systems may be explained in a similar way. For example, such a suppression would explain how aeciospores of flax rust (*Melampsora lini*) may form on a host variety that they are unable to reinfect (see page 67). I also refer, on page 102, to an example in Hessian fly where the presence of one virulent larva protects immediately adjacent avirulent larvae in the leaf base of a resistant wheat.

It would be satisfying to find that the same interactions associated with resistance and susceptibility occur in tuber tissue and in potato leaves. However, neither rishitin nor phytuberin was found either in uninoculated potato leaves or in leaves that had been inoculated with virulent or avirulent races of Phytophthora (Shih and Kuć reported in Kuć 1972). Potato leaves contain quantities of steroid glycoalkaloids (α-chaconine and α-so-

lanine) that are more than sufficient to prevent the growth of *P. infestans.*
Since these leaves support vigorous growth of virulent races, the glycoalkaloids are presumably present in a nontoxic or nonavailable form or are
localized away from infection sites within the cell. Whether they play a
role in hypersensitive resistance of leaf tissue is not known. These glycoalkaloids are also present in potato tuber peel and may accumulate near the
cut surface of uninoculated tuber tissue slices. However, in tubers inoculated with an avirulent race, synthesis of rishitin and phytuberin is stimulated (Kuć 1972).

A further complication is that even though petiole cells are rapidly
killed, the growth rate of avirulent intracellular hyphae is unchanged
during the first 2 or 3 hours of interaction (Tomiyama 1967). This must
mean that the substances responsible for host cell death do not inhibit
hyphal growth during this time, and that antifungal materials do not accumulate to an inhibitory level until later. In tuber tissue, rishitin is not detected until at least 12 hours after inoculation.

Pitt and Coombes (1969) have claimed that lysosomelike particles occur
in potato tuber tissue, and that they release hydrolytic enzymes in response
to cell injury by fungal pathogens. Their observations were limited to compatible host-pathogen combinations 3 or 4 days after infection. The rapid
release of lysosomal enzymes in incompatible combinations of potato and
P. infestans could well result in the early death of host cells. Tomiyama
(1971) has described cytoplasmic particles, released in petiole cells within
minutes of either infection with an incompatible strain or injury, that could
result from lysosome activity.

We might suppose that an avirulent pathogen prompts immediate
release of lysosome contents killing the host cell but having no immediate
effect on the pathogen. The released lysosome contents may then stimulate
synthesis of phytoalexins in uninvaded surrounding tissue. Wilson (1973)
recently reviewed the evidence for the role of lyosomes in plant-pathogen
interactions.

CORN

Another example of an antibiotic substance present in a nontoxic form
is the glucoside of the cyclic hydroxamate 2,4-dihydroxy-7-methoxy-1,4-
benzoxazin-3-one (DIMBOA), which is found in leaves of most normal
varieties of corn (*Zea mays*) and also of other cereals. When leaves
containing this glucoside are wounded, an enzymatic reaction releases the
aglucone DIMBOA. This is not only toxic to several fungi, including the
corn pathogen *Helminthosporium turcicum,* but also to larvae of the European corn borer *Ostrinia nubilalis* (Anglade and Molot 1967). A role of
cyclic hydroxamates in resistance to *H. turcicum* was recently suggested

by Couture et al. (1971), who made use of a recessive corn mutant (*bx*) in which the glucoside is present only in low concentrations (Hamilton 1964). The presence of the cyclic hydroxamate is readily scored when a blue coloration develops after crushing the mesocotyl of a 6-day-old seedling in 0.1M FeCl$_3$.

Hilu and Hooker (1963) had earlier shown that corn plants carrying the dominant gene *Ht* for resistance to *H. turcicum* produce greatly reduced chlorotic lesions when inoculated with an avirulent race (no races virulent on *Ht* plants have yet been found). Couture and his colleagues compared lesion areas on leaves of four kinds of corn inoculated with *H. turcicum*. The lines used were with (*BxBx*) and without (*bxbx*) normal levels of cyclic hydroxamates, and with (*HtHt*) and without (*htht*) the gene for resistance.

The low levels of cyclic hydroxamates were correlated with an approximate doubling of the lesion area in both resistant and susceptible plants. The resistance due to *Ht*, which was expressed even when only low levels of cyclic hydroxamates were formed, suggests that these compounds are only a first line of defense against *H. turcicum*, and that the mechanism controlled by *Ht* comes into play later. Corn plants with the genotype *bxbx* are extremely susceptible to aphid attack in the greenhouse (Couture: personal communication), suggesting that cyclic hydroxamates may be important to resistance to a range of insect parasites.

LEGUMES

Müller (1956) was the first to make use of the fact that the seed cavities of fresh pea or bean pods offer the opportunity to place drops of spore suspensions in contact with living tissue without an intervening barrier of epidermis or cuticle. Substances diffuse out of and into the drop. The substances that accumulate in the drop can be analyzed and studied further. Cruickshank and Perrin (1960) showed that pea pod tissue in contact with a suspension of spores of *Monilinia fructicola* formed the phytoalexin called pisatin (Figure 5.1), and in 1963, that bean pod tissue treated in the same way formed another phytoalexin called phaseollin (Figure 5.1).

Both phytoalexins are produced in response to pathogens and nonpathogens of their respective hosts. Pisatin, stimulated by nonpathogens, accumulates to levels that are toxic to them; on the other hand, pea pathogens either do not stimulate levels sufficient to bring about resistance, or are relatively insensitive to it, or degrade it. Phaseollin is produced by bean pods and foliage and, like pisatin, shows selective toxicity to pathogens and nonpathogens of the source tissues. According to Rahe et al. (1969), phaseollin is important in determining varietal resistance to *Colletotrichum lindemuthianum*, the cause of anthracnose in beans. It accumulates most rapidly in bean hypocotyls inoculated with an avirulent race. Inoculation with a virulent race produces a much lower level of accumulation. Three races of the same organism examined by Bailey and Deverall (1971) were equally sensitive to phaseollin, suggesting that if phaseollin has a role in race-specific resistance, it would depend on the amount produced and not on differential toxicity.

Phytophthora megasperma var. *sojae* is a root parasite of soybeans (*Glycine max*), which can be controlled by oligogenic race-specific resistance. Keen et al. (1971) set out to confirm earlier reports that resistance resulted from the production of a phytoalexin. They showed that 6α-hydroxyphaseollin (HP) (Figure 5.1) was formed in hypocotyls of seedlings of the resistant variety Harosoy 63 within two or three days after inoculation with mycelium of an avirulent race of *P. megasperma sojae*. The near-isogenic susceptible variety Harosoy, inoculated with the same race, also produced HP but at rates that were at least 10 times lower.

Frank and Paxton (1970), working with the same system, claimed that an unidentified fluorescent compound was the major phytoalexin, and that this was induced to accumulate in hypocotyl tissue of Harosoy 63 by a low molecular weight glycoprotein, isolated from culture filtrates of *P. megasperma* var. *sojae* (Frank and Paxton 1971).

Keen and Horsch (1972) have claimed that the differential production of HP in response to the avirulent pathogen was only clearly shown by seedling hypocotyls. Roots, pods, and tissue culture callus of the two varieties produced similar amounts of HP. Evidently the site of action of the gene for resistance in Harosoy 63 is limited to hypocotyl tissue.

SPECIFIC INDUCTION OF PHYTOALEXINS

In the search for an explanation of specificity, much attention has been focused on how phytoalexins are induced. Some are produced in quantity in response to wounding, whereas others are only produced after inoculation with either a nonpathogen or an avirulent race of a pathogen. At first sight, those in the last category appear to be better candidates for investigating specificity, but this is not necessarily so. Phytoalexins in the first category may well accumulate because an avirulent pathogen "wounds" the host and a virulent pathogen does not.

Pisatin and phaseollin are probably the two phytoalexins whose induction has been most intensively studied. An almost bewildering variety of compounds induce pisatin formation in drop diffusion tests in pea pods. Hadwiger and Schwochau (1969) proposed that the synthesis of pisatin is repressed in normal tissue and, following the Jacob-Monod model for regulation, suggested that synthesis was the result of derepression. This led to a scrutiny of pisatin inducers to see whether their known properties would explain their role as derepressors. Since pisatin is an end point in a synthetic pathway, and it is likely that all the enzymes in that pathway are regulated coordinately, particular attention was paid to the key step in aromatic biosynthesis controlled by phenylalanine ammonia lyase (PAL). This enzyme occurs at a branch point, diverting material from protein synthesis to phenylpropanoid metabolism (flavonoids, lignin, lignanes, certain alkaloids, and conjugates such as chlorogenic acid). All the nearly 100 compounds that increased pisatin accumulation also increased PAL activity, although the reverse was not true of several tested. Pisatin inducers included heavy metals, polypeptides, and antibiotics and drugs which bind with DNA. Among the latter were a number of compounds with 3 planar rings, which are known to intercalate in DNA (Hadwiger and Schwochau 1971).

Unfortunately, no very clear picture of how these inducers work has yet emerged. Although they may interfere with a repressor or promote transcription by producing conformational changes in host DNA, they would be expected to do this for many other regulated pathways. The lack of specificity leaves me with the feeling that anything mildly injurious to the cell that does not prevent RNA and protein synthesis may have the same effect. Perhaps none of these substances directly regulates PAL activity, but they work instead by disturbing a sensitive network. The resulting imbalance leads to the derepression of PAL and formation of pisatin and probably many other compounds. So far no inducer substance that is clearly a product of a gene for avirulence in a pathogen has been characterized. Neither has any product of a specific gene for resistance yet been isolated that can be shown to have a complementary role to an inducer.

DEGRADATION AND DETOXIFICATION

If host plant resistance is based on phytoalexin production, it will fail if the parasite is either insensitive to the phytoalexin or does not induce its formation. Higgins and Millar (1969a) showed that leaves of alfalfa (*Medicago sativa*) form a phytoalexin following inoculation with spores of *Stemphylium botryosum,* a pathogen, or with spores of *Helminthosporium turcicum,* a nonpathogen of alfalfa. Filtrates from suspensions of germinating spores of both fungi caused accumulation of the phytoalexin, but the presence of germinated *S. botryosum* spores prevented its accumulation, whereas *H. turcicum* spores did not. These authors (1969b) showed that *S. botryosum* degraded the phytoalexin, induced either by its own filtrates or by spores of *H. turcicum,* to form two phenolic compounds that were not inhibitory to germ tube growth of either fungus. A further proof of the role of phytoalexin degradation in the pathogenicity of *S.*

botryosum to alfalfa would be the demonstration that mutants unable to degrade it are nonpathogenic, or that resistance in alfalfa could result from the production of an altered phytoalexin that *S. botryosum* was unable to degrade.

Pisatin is also degraded by pea pathogens, and DeWit-Elshove and Fuchs (1971) have shown that pisatin degradation may be subject to catabolite repression. High glucose or sucrose strongly inhibit the breakdown of ^{14}C-labeled pisatin added to cultures of *Fusarium oxysporum pisi* and *F. solani pisi.*

Insensitivity to a phytoalexin can arise from the presence of interfering substances. Mansfield and Deverall (1971) showed that extracts of pollen of the broad bean *Vicia faba* render spores of *Botrytis cinerea* insensitive to wyerone acid, an antifungal product of infected bean tissues. Their findings explain field observations that the presence of pollen leads to greatly increased lesion size on beans.

Detoxification mechanisms that deal with preformed defense compounds, and presumably phytoalexins as well, appear to be common among insects. They are often responsible for insecticide resistance. A good example comes from a comparison of midgut microsomal oxidase activities of lepidopterous larvae. Krieger et al. (1971) examined 35 species and placed them in 3 groups according to the number of plant families that serve as larval food plants. The first group included monophagous species, normally confined to plants of one family, or two closely related families. Species in the second group fed on from 2 to 10 plant families, and those in the third were polyphagous, feeding on more than 10 plant families. The microsomal oxidase activities were estimated by measuring the rate at which aldrin was converted to its epoxide dieldrin by homogenized gut preparations. The mean activity for 12 polyphagous species was nearly 15 times greater than the mean of 8 monophagous species, but the mean of 15 species in the second group was only 4 times greater than that of the monophagous group. The authors point out that polyphagous insects are equipped to deal with a greater range of natural pesticides evolved by plants to protect themselves against insect attack.

Miles (1969) suggested that the salivary phenolases of some Hemiptera counter the defensive phenolic toxins produced by their host plants by speeding their oxidation to nontoxic end products. There are comparable detoxifications among plant pathogens; for example, the cell-bound dehydrogenases of the soft rot bacterium (*Erwinia carotovora*) reduce toxic quinones, resulting from host phenol oxidase activity, that limit bacterial growth in potato tuber tissue (Lovrekovich et al. 1967). However, this has not yet been related to the relative virulence or avirulence of isolates of *Erwinia* in host tissue.

Some Other Insect Resistance Mechanisms

Some plants are resistant to insects because they lack an attractant or produce a repellant (Hsiao 1969). The tetracyclic triterpenoids, or cucurbitacins, responsible for the bitter flavor of some cucumbers, on the one hand, are specific feeding attractants for cucumber beetles (Crysolimadeae), but on the other hand, protect plants from attack by two-spotted mites *Tetranychus urticae* (DaCosta and Jones 1971). The presence of cucurbitacins is determined by the gene *Bi*. DaCosta and Jones argued that once the beetles were adapted to overcome cucurbitacins, they evolved the ability to respond to them as attractants because of the selective advantages of having a food source on which they alone could feed with impunity.

Hovanitz (1969) claimed that strains of the cabbage butterfly *Pieris rapae* could be induced to prefer host plants with different concentrations of mustard oil (allyl isothiocyanate) by conditioning for several generations. F_1's between strains conditioned on Kale (10^{-8}M allyl isothiocyanate) and mustard (10^{-6}M) showed a preference for an intermediate concentration (10^{-7}M) of mustard oil.

Plants may also be unpalatable to insects if they form products that interfere with insect digestion. Green and Ryan (1972) noted that potato and tomato tissues accumulated proteinase inhibitors after mechanical and insect injury. They suggested that these inhibitors would interfere with proteinases in the insect digestive tract and thus act as insect deterrents.

Other Effects of Parasites

HISTOLOGY AND ULTRASTRUCTURE

In this chapter we are interested in the differences between resistant and susceptible interactions and what they can tell us about the functions of the genes that determine those differences. Examination of fresh material, stained sections, or cleared and stained tissues with the light microscope reveals the route of parasite invasion, its extent, the kind of damage, and other useful information, but it tells us little about the involvement of cell organelles, membranes, and submicroscopic structures. Electron microscopy of thin sections and freeze-etched sections reveals much more of this detail.

In many obligate parasitic fungi and a few nonobligate parasites, such as *Phytophthora infestans*, the most intimate contact with host tissue occurs through haustoria. Haustoria are essentially specialized branches of the parasite mycelium which develop from a mother cell appressed to the host cell. The developing haustorium penetrates the host cell wall by means of an infection peg. Once through the cell wall, the host cell plasma membrane invaginates around the developing haustorium, which swells out into the host cell vacuole from the relatively narrow neck through the host cell wall (Figure 5.2). Haustoria generally contain one or several nuclei and function as absorptive organs (Figure 5.3). Their structure and development was recently reviewed by Ehrlich and Ehrlich (1971).

Where host resistance permits little or no development of parasite mycelium, haustorial penetration may lead to rapid death of the host cell. This is characteristic of the resistant reactions of the flax variety Ottawa 770B to flax rust (Littlefield and Aronson 1969) or the cowpea (*Vigna sinensis*) variety Queen Ann to cowpea rust (*Uromyces phaseoli vignae*) (Heath and Heath 1971) (Figure 5.2:C). Both these examples also show cases where individual penetrating haustoria become enclosed in a thick sheath, which appears to seal it off, and the host cell does not become necrotic (Figure 5.2:B). Where resistant and susceptible reactions are less extreme, there are few if any differences observable at the ultrastructural level during the early stages of infection. The differences observed in later stages are quantitative, rather than qualitative, and involve extent of development, growth rate, and the speed at which senescence occurs.

The ultrastructural effects of phytotoxins have also been investigated. The HV-toxin of *Helminthosporium victoriae* produces blisterlike invaginations of the plasma membrane, which supports the suggestion that toxin activity affects membrane function (Goodman 1972).

CULTURE FILTRATES

The search for phytoalexin inducers has naturally included pathogen metabolites. Some of these are also known to have other effects. The host-specific toxins of *Helminthosporium* spp. are cases in point. In Holland, Kaars Sijpsteijn and her colleagues have used two host-parasite systems to examine the effects of pathogen culture filtrates on host tissue. Unfortunately, the effect of filtrates of avirulent cultures of *Venturia inaequalis* in producing wilting symptoms in detached apple leaves reported by Raa (1968) could not be repeated by Boone (1971) or Kuć (1972). However, Pellizzari et al. (1970) reported a doubling of the rate of electrolyte

FIGURE 5.2 *(facing page)*
Reactions of cowpea *Vigna sinensis* to haustoria of the rust *Uromyces phaseoli vignae*. A—susceptible reaction 6 days after inoculation ($\times$ 7,280). B—immune reaction showing sheathing response ($\times$ 5,950). C—immune reaction showing necrotic response ($\times$ 7,280). (B, C fixed 28 hours after inoculation.) (From Heath and Heath 1971.)

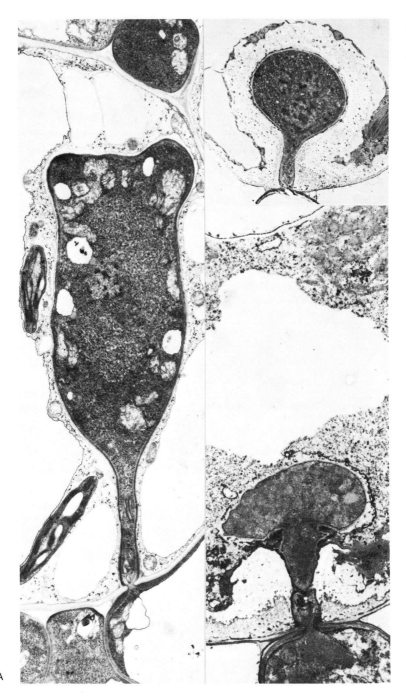

A

B

C

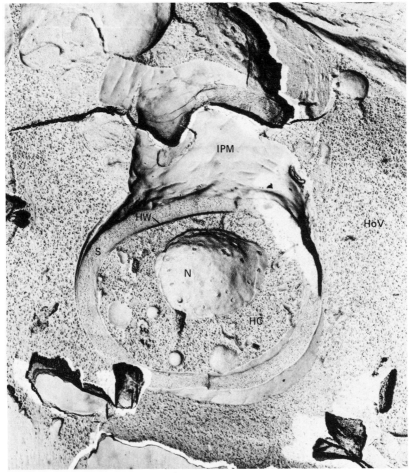

FIGURE 5.3

Freeze-etched cross-fracture of a haustorium of *Melampsora lini* in a susceptible host cell of flax (× 14,100). HC—haustorial cytoplasm, HW—haustorial wall, HoV—host vacuole, IPM—invaginated host plasma membrane, N—haustorial nucleus surface view, S—haustorial sheath. (From Littlefield and Bracker 1972.)

leakage from detached apple leaves, showing a hypersensitive response to inoculation with a spore suspension.

In other experiments, van Dijkman and Kaars Sijpsteijn (1971) reported that compounds present in culture filtrates of avirulent races of *Cladosporium fulvum* increased the rate of leakage of ^{32}P from labeled leaf discs of resistant tomato varieties. In several interactions involving the genes Cf_1 and Cf_2, and races 0, 1, 2, 1.2, and 1.2.4, isotope leakage only exceeded the control value in resistant or avirulent interactions. Although phytoalexins that are toxic to *C. fulvum* have not been reported in tomatoes, it may be that effects on host cell permeability are instrumental in the induction of defense reactions. On the other hand, the effects of culture filtrates could be spurious and not related to the induction of specific resistance.

A TEMPERATURE SENSITIVE SYSTEM

Comparisons between resistant and susceptible interactions usually involve altering either host or parasite genotypes. Thus a resistant interaction can, in theory, be converted to a susceptible interaction either by substituting an allele for susceptibility in the host or by substituting an allele for virulence in the pathogen. In practice, the substitution is difficult to carry out so that the two host lines or the two pathogen cultures are isogenic (have identical genotypes except for the particular locus being studied). If they are not isogenic, the differences observed may be due to other loci. In the potato late blight system, neither host lines nor parasite cultures are isogenic, and we therefore do not know whether the observed differences are due primarily to single genes (although the gene-for-gene hypothesis suggests that they are). In wheat a series of near-isogenic lines for stem rust resistance have minimized this difficulty (Loegering and Harmon 1969). Temperature-sensitive genes provide another method of overcoming it. The *Sr6* gene for resistance to stem rust from the wheat variety Red Egyptian is temperature sensitive (Loegering and Geiss 1957). At 20°C its reaction type is 0; (resistant) to an avirulent race but at 24–25°C the same race gives a 4 (susceptible) reaction. Since wheat is inbred, it is relatively simple to produce stable homozygous lines that appear to be uniform for disease reaction. The genetic stability of the *Puccinia* culture was assured by storing uredospores in liquid nitrogen. Daly and others have used the *Sr6* system to compare the biochemical changes that occur, with the same combination of host and pathogen, at both temperatures. As a further control, Daly made use of a near-isogenic rust susceptible line homozygous for the recessive allele *sr6*. Over a period of several years, changes in respiration (Antonelli and Daly 1966), tissue phenolic content (Seevers and Daly 1970a), and indoleacetic acid decarboxylation (Antonelli and Daly 1966), and peroxidase activities (Seevers and Daly 1970b; Seevers et al. 1971) were studied with a view to establishing the nature of resistance conditioned by *Sr6*.

The decision to examine the enzyme activities associated with these systems was based on earlier observations that oxidative processes of several kinds are more active in resistant host-parasite interactions than in susceptible host-parasite interactions (Allen 1959).

The results are of considerable interest since the study was designed to reveal the effects of a single resistance gene. Contrary to expectation, during the first 5 days no differences in respiration activity were observed between near-isogenic lines with *Sr6* or *sr6* at 20°C. Both rose in parallel. The later rise in respiration of susceptible interactions was presumed to result from the energy demands for sporulation by the pathogen, an idea supported by a fall in C_6/C_1 ratios from 0.54 to 0.38 that occurred after day 5. C_6/C_1 ratios are determined by supplying tissues with radioactive glucose-^{14}C labeled at either the C-1 or C-6 positions. The relative rates at which ^{14}C appears in respired CO_2 from the two kinds of labeled glucose give an estimate of the relative contributions of the glycolytic pathway, in which C-1 and C-6 contribute equally to respired CO_2, and of the pentose phosphate pathway, in which CO_2 is derived only from the C-1 of glucose. Indoleacetic acid (IAA) decarboxylation activity increased in resistant interactions (*Sr6* at 20°C) but not until 3 days after infection. By the sixth day the activity was 4 to 10 times that in healthy controls, whereas the activity in susceptible interactions was lower than the controls.

A rise in peroxidase activity in the resistant interaction was correlated with the increase in IAA decarboxylation activity and could account for it. No changes in concentration of phenols produced during infection of *Sr6* and *sr6* lines were detected. Experiments in which infected leaves were transferred from 20°C to 26°C at various times after infection showed that the critical period for establishing a resistant or a susceptible pattern occurred about 3 days after inoculation. Several experiments of this kind failed to show that any of the observed biochemical changes were crucial. For example, infected leaves of *Sr6,* kept at 20°C for 6 days, developed an increased peroxidase level, and when they were transferred to 26°C, peroxidase activity did not decline significantly, although the reaction type changed from 0; to 3–4. Furthermore, the increased peroxidase activity was attributable to only one band among 14 identified by electrophoresis on acrylamide gel (Seevers et al. 1971). Comparable experiments with another gene for rust resistance (*Sr11*) showed that increased peroxidase activity was associated with the same isozyme band (Daly et al. 1971). The gene *Sr11* is clearly different from *Sr6* since it is not temperature sensitive and does not respond to ethylene (see below). We may conclude that increased peroxidase activity is a secondary effect of resistance and not a determinant.

DIFFERENT GENES—DIFFERENT EFFECTS

The examples of different resistance genes in the same host that we have examined so far have given some indication of different responses to environmental stimuli. For example, *Sr6* in wheat is sensitive to temperature and ethylene, whereas *Srl1* is sensitive to neither. We might ask whether they interact with an avirulent culture in the same way, or if the initial interactions are different, whether the subsequent steps are similar, and if so, to what extent. So far no intrinsic differences in their resistant phenotypes have been found. For other stem rust resistance genes in wheat, somewhat larger differences exist. Some allow greater development of infection (reaction types 1 or 2) or determine a mesothetic reaction, but these have not yet been analyzed for differences in their biochemistry. In potatoes there are few, if any, phenotypic differences among genes for race-specific resistance to blight. In any case our most detailed observations are based on only one—the gene *R1*.

Working with the powdery mildews of barley and wheat, Ellingboe (1972) and his colleagues have laid a foundation for biochemistry by defining the points during the early stages of infection at which resistance is expressed. Several different genes were examined in each host. Populations of conidia, fairly well synchronized for germination and subsequent development, were applied to leaves of appropriate host genotypes under optimal conditions for infection. In no case did resistance interfere with spore germination, the formation of mature appressoria, or formation of small infection pegs on the appressoria. It was expressed only after the host and the parasite had made intimate, membrane-to-membrane contact. Resistance was never 100 percent effective in inhibiting development of haustoria and secondary hyphae. The number of conidia that progressed beyond that stage and the extent of their development was a characteristic of the resistance gene concerned. The results are summarized in Figure 5.4.

Slesinski and Ellingboe (1971) also compared the amounts of ^{32}S transferred from labeled wheat leaves to the pathogen in compatible and incompatible interactions. They applied a parlodion solution to the infected surfaces of labeled leaves, at different times after inoculation. The dried parlodion was stripped off, as a film containing the embedded parasite, for radioactivity determination in a scintillation counter. As might be expected, the amount of label transferred was generally proportional to the extent of parasite development in the host. In the *P2/Pm2* incompatible interaction, label transfer in the first 26 hours was greater than in the compatible control. Whether this was due to a greater growth of the parasite or to greater physiological activity was not determined. In a comparison of amounts of label transferred in the three compatible interactions *P1/pm1, p1/pm1,* and *p1/Pm1,* the least amount of label was transferred in *p1/Pm1*. If this was not an effect of variation in genetic background, it suggests that although all three compatible interactions are macroscopically identical, they may differ physiologically. One possibility

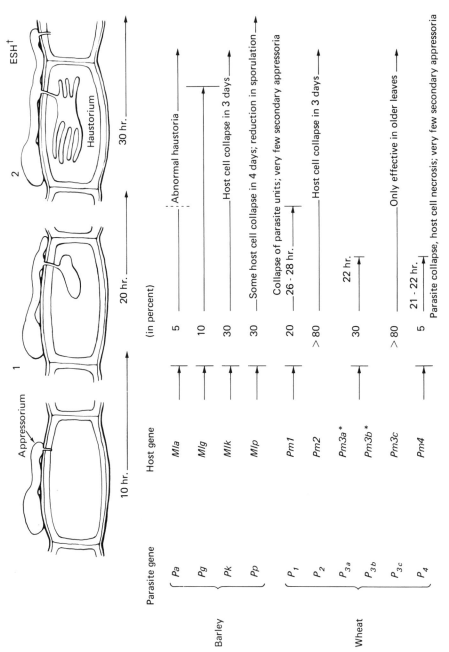

FIGURE 5.4
Diagramatic representation of points at which resistance genes in barley and wheat block the early development of powdery mildew infections. (Based on Ellingboe 1972.)

is that *Pml* has a pleiotropic effect which limits isotope transfer in some other way than in the incompatible interaction *Pl/Pml*.

SUSCEPTIBILITY AS AN INDUCED RESPONSE

The evidence that resistance is due principally to the production of toxic chemicals that limit pathogen growth, as we have seen, is not entirely convincing. The antibiotic properties of phytoalexins and the circumstances under which they arise are not disputed, but that they are the only or even the primary reason why pathogen growth is inhibited can be disputed. Daly et al. (1971) found that wheat leaves carrying the resistance gene *Sr6* became susceptible to stem rust at 20°C when treated with 75 ppm ethylene after inoculation. This result was unexpected. Ethylene increases peroxidase activity in plant tissues, and this activity was known to be greater in infected resistant plants than in infected near-isogenic susceptible plants. Higher peroxidase levels were found in leaves after ethylene treatment, and this might have been expected to result in increased aromatic biosynthesis of antimicrobial compounds, but in fact a susceptible reaction was obtained. Although higher peroxidase levels were induced in leaves of plants carrying *Sr11*, resistance was unaltered. The role of ethylene in inducing susceptibility in *Sr6* plants suggested to these authors the possibility that susceptibility is normally induced, and that a series of adjustments are made in host metabolism to accommodate the pathogen. If these adjustments are not made, then injury or death of the pathogen may in turn result in other changes in host metabolism, which we have so far referred to as defense reactions. Zucker (1968, 1971) found that potato tuber tissue and *Xanthium* leaves both contain mechanisms, dependent on protein synthesis for their expression, that destroy PAL activity, and he has discussed their role in getting rid of unwanted enzyme activities in higher plant cells. Susceptibility might well result from induction of similar host cell proteolytic enzymes or inhibitors that prevent the development of a normal defense response.

The evidence from sequential inoculations of potato tuber tissue with *Phytophthora infestans* suggests that both mechanisms may operate in the same system. If an avirulent inoculum is applied first, an incompatible reaction is established that is not reversed by virulent inoculum applied later. If virulent inoculum is applied first, a compatible reaction is established and is also unchanged following challenge with an avirulent inoculum applied later.

In the barley-powdery mildew system, Dyer and Scott (1972) compared the compatible interaction *Pa/mla* with the incompatible interaction *Pa/ Mla* (see Figure 5.2 for explanation of symbols). The resistant and susceptible barley lines were near-isogenic. Whole leaves were homogenized 24 hours after inoculation and their polysome contents compared, following sucrose density gradient centrifugation. In the compatible (susceptible) interaction, chloroplast polysomes had been lost, whereas no change was observed in the incompatible (resistant) interaction. Since the mildew fungus invades the epidermis and has only begun to form haustoria by 24 hours, Dyer and Scott suggested that the change they observed in the susceptible host was induced by a product that diffused from the pathogen. The critical reactions that determine subsequent development occur during the period 20 to 24 hours after inoculation. If these observations are correct and the difference is not simply due to failure of the mildew to penetrate resistant cells, the diffusing substance could be an inducer of susceptibility.

A MODEL BASED ON GENE REGULATION

The changes in host tissue associated with resistance or susceptibility are clearly the result of genetic regulation and differentiation. Since these processes are still poorly understood in healthy tissues, it is not surprising that progress in understanding the nature of host-parasite interaction is slow. A simple model for gene regulation was advanced by Britten and Davidson (1969), and although it is still theoretical, a consideration of how it might work in host-parasite interaction will show the sort of problems that need to be solved. The Britten and Davidson model assigns the roles of regulation and integration of cell metabolism and differentiation to the large numbers of repetitive nucleotide sequences in the genomes of higher organisms. This redundant DNA controls the responses of large batteries of genes to specific inducing agents. The model requires five genetic elements: (1) A sensor gene or nucleotide sequence which binds agents that induce a specific pattern of activity in the genome. Binding results in the activation of a linked gene, or genes, called (2) an integrator that synthesizes (3) an activator RNA. The activator RNA forms specific complexes with (4) receptor genes that are linked to (5) producer genes causing them to be transcribed, probably by relieving histone-mediated repression. Producer genes form all RNA's other than those involved in regulation of the genome. For example, they may code the messenger RNA for synthesis of a polypeptide or code for a transfer RNA. Redundancy occurs either because many different sensor genes have common integrator sequences that enable them to regulate the same producer genes, or because many different producer genes have common receptor sequences that enable them to respond to the same activator RNA. A battery of producer genes are controlled together when a particular sensor gene activates its set of integrator genes. The degree of physiological coordination required will determine whether the redundancy occurs at the level of integrator sequences or receptor sequences or a combination of the two.

Figure 5.5 gives a simple example of how these elements might function in relation to race-specific resistance. The resistance genes *R1, R2,* and *R3* are shown as sensor genes regulating different combinations of producer genes, whose activation results in the expression of resistance. Although this can account for the phenotypic differences that are often observed among different resistant or avirulent interactions, it seems likely that such batteries will have many producer genes in common. If the sensor genes are allelic, one possibility is that all the alleles at a specific locus will regulate the same battery of producer genes so that all phenotypes are the

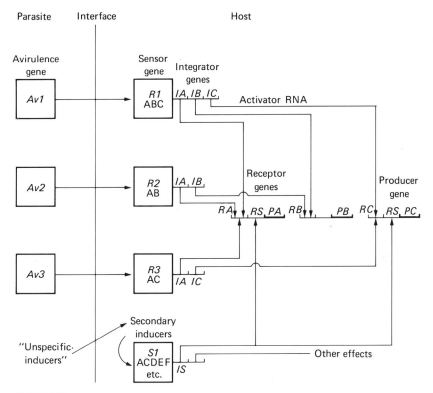

FIGURE 5.5

A model for host-parasite interaction. The sensor genes R1, R2, and R3 are resistance genes in the host that bind with the product of a specific avirulence gene from the parasite. The synthesis and diffusion of activator RNA to receptor genes results in the activation of their linked producer genes, producing the changes in metabolism associated with resistance. Sensor gene S1 represents the class of sensor that responds to unspecific inducers. (After Britten and Davidson 1969.)

same. It is also possible that different alleles may be associated with different integrator genes, so that alleles at the same locus may regulate different producer genes and thus also have different resistance phenotypes. Although examples of both kinds are known among multiple allelic resistance genes, it is difficult to be sure that dissimilar phenotypes are not the result of other, uncontrolled, genetic differences. If each resistance gene regulates a large battery of producer genes, it may be very difficult to identify any one particular end product as playing the major role in disease resistance. This is certainly born out by experience so far.

In this model specificity seems most likely to occur at the level of the sensor genes. How then can we interpret the data from genetic studies in corn and flax on intragenic recombination? (See pages 22–26.) Each allele for resistance is a nucleotide sequence that determines a unique binding specificity. These sequences will recombine by crossing-over, and providing they do not interfere with each other, will produce recombinant specificities that bind with either or both inducers. The example found by Shepherd and Mayo (1972), where the cross of two alleles ($L^2 \times L^{10}$) did not produce a functional recombinant specificity, can be explained if the two different binding specificities cannot be accommodated by the sequence of nucleotides available without interfering with each other. The recombinant sequence may give a new binding specificity that would only respond to the inducer product of a different avirulence locus of the rust pathogen, or alternatively, it may be nonfunctional. It is equally possible that recombination involving the integrator genes linked with L^2 and L^{10} could lead to a situation where their transcription was either blocked or defective. Or if transcription did occur, a nonfunctional activator RNA was produced. This last suggestion is probably closest to the explanation advanced by Shepherd and Mayo (page 26).

It was mentioned earlier that no materials that could be the products of specific avirulence genes have yet been identified in spite of intensive searches. They may be labile proteins or even RNA sequences that have special functions in parasite metabolism. On the other hand, they must diffuse readily from the intact pathogen into the host cell nucleus if they are to perform their function according to the model.

Unspecific inducers such as substances from nonpathogens, boiled extracts of *Phytophthora infestans,* or the many compounds that elicit pisatin formation probably do not act directly on "resistance" sensor genes. They might act as irritants which provoke the formation of secondary inducers detected by other sensors such as S1 in Figure 5.5. The result would be induction of responses that may be similar in some respects, but different in others, to the resistant response.

Although Figure 5.5 deals only with resistance as an induced response, it is not difficult to account for induced susceptibility by a similar set of controls. The model is also entirely consistent with the observation that once either a resistant or a susceptible response has been induced, and allowed to progress for several hours so that the host cell is committed, it will not normally be reversible. If induced resistance and induced susceptibility both occur in the same system, their competitive interaction must normally be resolved in one direction or the other. Although it is possible that it is generally resolved in favor of the expression of resistance, this may not always be so. Indeed, if reactions in either direction are equally likely for each local interaction, we have a ready-made explanation for mesothetic reactions.

Normally I would expect that a virulent pathogen would produce materials that induce susceptibility and not materials that induce resistance. Whether any given pair of alleles for virulence or avirulence in a pathogen involves one or the other may be difficult to decide. It is tempting to say that dominant avirulence implies production of a substance that induces resistance and has the effect of overriding inducers of susceptibility that may be present. In this case recessive virulence merely implies absence of the substance that induces resistance. Resistance in the host is more often dominant than recessive, but the model is not particularly helpful in explaining why this should be so.

THE INTERACTION IS LOCALIZED

When resistance is not preformed but develops in response to parasite invasion, we might expect that it would be localized. The response is generated at the site of interaction and should involve only the immediately neighboring cells. Any wider involvement of other tissues could, if carried to extremes, result in a catastrophic hypersensitive reaction destroying the organ or even the whole plant. Potato varieties extremely susceptible to virus X develop acute necrosis upon infection. In practice, these varieties are either healthy or killed outright (Smith 1957), so that the virus is self-eliminating. Virus X persists on varieties that are not as susceptible.

There are several ways of demonstrating that resistance is localized. One is by means of preinoculation. For example, flax carrying the gene L for resistance to rust was partially resistant to a virulent race if inoculated at least 4 hours previously with an avirulent race (Littlefield 1969). When either distal or proximal halves of leaves were preinoculated with the avirulent race, infection and development of the virulent race was normal on the adjacent untreated halves. Although the mechanism of induced

A

FIGURE 5.6

Localization of resistance in wheat chimaeras. A—three leaves inoculated with one race of
Erysiphe graminis tritici, showing, from left to right, resistant and susceptible interactions.
B—three leaves inoculated with one race of *Puccinia graminis tritici*, showing from left to
right, a range of type 2 (resistant) pustules and type 3+ (susceptible) pustules. In each case
the center leaf is a chimaera. (A from McIntosh and Baker 1969, and B from Baker:
unpublished.)

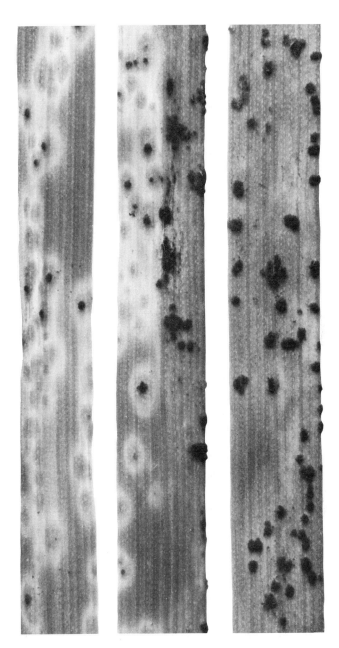

B

resistance was partly explained by competition for stomatal entry sites, the size and rate of development of the pustules that were formed were significantly reduced. This points to a localized physiological resistance mechanism. Induction appeared to be nonspecific since nonpathogens of flax *(Puccinia graminis tritici* and *P. recondita tritici)* had similar effects. Mechanical blocking of infection sites can also account for the resistance to stem rust in wheat induced by preinoculation reported by Cheung and Barber (1972).

The most satisfactory method for demonstrating localized resistance is to use chimaeras in which two adjacent tissues have different genotypes and therefore different reactions to a race of a pathogen. Chimaeras are difficult to produce experimentally, but two chance ones observed in wheat by McIntosh and Baker (1969) and Baker (unpublished) are of interest.

Assigning genes to chromosomes in wheat is usually carried out with the aid of a set of lines of the variety Chinese Spring, originally developed by E. R. Sears. Each line is monosomic for one of the 21 pairs of chromosomes. McIntosh and Baker found that the gene *Pm3a* for resistance to powdery mildew (*Erysiphe graminis tritici*) in Asosan wheat was carried by chromosome 1A. This was shown by the normal 3:1 segregation for resistance in F_2's from crosses with all the monosomic stocks except that involving chromosome 1A. In this progeny only 15 of 239 plants were susceptible, presumably because they were nullisomic. Three of the resistant plants were chimaeras, having some leaves with adjacent sectors of resistant and susceptible tissue. The susceptible sectors were assumed to result from the loss, in an early embryonic stage, of the Asosan 1A chromosome carrying *Pm3a*. In all 3 the longitudinal dividing line between the tissues occurred very close to the leaf midrib. An example is shown in Figure 5.6:A.

A second example (Figure 5.6:B) was found by Baker (unpublished) in an aneuploid wheat (2n = 43) in which resistance to stem rust *(P. graminis trititici)* was carried by a single chromosome from *Agropyron elongatum* in a background genotype of the variety Gabo. The resistant tissues show a range of resistant infection types; those permitting sporulation would be classed as "2," whereas the susceptible tissues show "3+" infection types. In the chimaera the susceptible leaf sector presumably lost the Agropyron univalent through misdivision.

In both chimaeras the clear demarcation lines between resistant and susceptible tissues revealed by the pathogens show that both resistance and susceptibility are localized.

Burrows (1970) reported a similar example in which resistance to crown rust in oats was limited to those tissues in leaf chimaeras that carried the resistance gene *Pc-15*.

MUTATIONS AFFECTING RESISTANCE

In Chapter 2, I discussed some work on induced mutation to resistance in crop plants. We must now ask what this information and comparable experiments to induce virulent mutants in fungal pathogens can tell us about the nature of the interaction.

First, we might compare induced resistant mutants with naturally occurring resistant forms. The very practical question arises, are induced resistant mutants also race specific?

Two problems arise that make this a difficult question to answer. First, the duplication of a natural gene for resistance may be due to contamination. Unless this can be ruled out by adequate controls or tests (see page 18), the data are of little interest. The second problem is that the mutants that satisfy this criterion have generally not been tested or used on a sufficiently large scale for us to be sure they are not race specific. The *ml-o* mutant for mildew resistance in barley is a good example. Plants homozygous for this recessive mutant are not vigorous and have low yields, probably because of the necrotic flecking. If *ml-o* arose in nature, it would almost certainly fail to survive in the absence of heavy mildew infections. This type of induced resistance has little or no practical value if it compromises yield in the absence of the parasite. In a recent article on the genetics of barley mildew, Wolfe (1972) noted evidence suggesting that single recessive alleles for resistance in four barley lines of Ethiopian origin are all located at or near *ml-o*.

The problem of necrotic lesions associated with resistance can be circumvented. Langford (1948) showed that the dominant gene Cf_2 for resistance to *Cladosporium fulvum*, isolated from the currant tomato *(Lycopersicon pimpinellifolium)*, would produce a severe autogenous necrosis in the cultivated tomato *(L. esculentum)* unless accompanied by an unlinked dominant gene *Ne* from *L. pimpinellifolium* that suppressed it. *Ne* had no effect on the expression of resistance to the fungus. Necrotic *(Cf$_2$-nene)* and nonnecrotic *(Cf$_2$-Ne-)* plants were equally resistant. Whether necrotic plants were also resistant to races of *C. fulvum* virulent on nonnecrotic *Cf$_2$-Ne-* plants was not recorded.

The report by Schwarzbach (in Wolfe 1972) that the association between necrosis and resistance due to *ml-o* may be broken suggests that a similar system may occur in barley.

Few studies have examined the nature of resistance in induced mutants. The oat mutants resistant to victorin or sorghum mutants resistant to *Periconia* toxin are among the more promising because of the ease with which they can be selected (page 19).

Induced mutation to susceptibility could throw light on the nature of resistance by providing a way of dissecting the resistant interaction. The regulation theory we considered earlier suggests that induced susceptible mutants of a monogenic resistant host might not be restricted to the sensor gene for resistance. Mutation of the regulated genes could well result in defects in the sequence of responses, so that key inhibitory materials were not formed. In practice, this could prove to be a more effective method for identifying them than those used now. The approach has much to recommend it. Many times geneticists have been able to unravel a system that they did not understand by inducing defective mutants that blocked it at different steps. If the blocks can be arranged in a logical sequence, they may reveal the operation of the undamaged system. Very few attempts have yet been made to induce susceptible mutants and systematically compare them in subsequent genetic analyses.

Mutation to Virulence

The selection of induced virulent mutants in pathogenic fungi has been attempted in at least seven different systems (Table 5.1), with varying degrees of success. Large numbers of fungal spores can be treated with a mutagen and mutations from avirulence to virulence screened using resistant host tissue. Genetic markers may be used to eliminate contaminants, and the efficiency of the selection method can be tested in reconstruction experiments in which avirulent and virulent spores are mixed in proportions that simulate mutation frequencies.

Insensitivity, or failure to detect "mutants" present in very low frequencies, may result from several causes. The great majority of viable spores in the inoculum are avirulent, and the resistant reactions that they provoke may interfere with the establishment of virulent infections. Even if avirulent spores, or adjacent avirulent cells in a multicellular spore, do not provoke interfering resistant reactions, they may still compete for, and block, available infection sites. If the probability of a single spore giving rise to an infection is low, the frequency of virulent mutants may never be high enough to ensure recovery of at least one lesion unless the scale of the experiment is so large as to be unmanageable.

It is also possible that the mutant, or virulent, phenotype of a spore may not be expressed in the time interval between germination and penetration and the germling will be eliminated by the host reaction that its original phenotype triggered. This is why mutations to self-compatibility in self-incompatible flowering plants generally cannot be recovered by treating mature pollen with a mutagen and applying it to an incompatible stigma (Lewis 1949). If mutagen treatment is given to flower buds prior to or during meiosis, the period of growth and development before pollination

TABLE 5.1
Induced mutation to virulence.

System	Agent	Spores	Success	Max. mutation rate	Reference
Melampsora–Linum	U.V.	n + n	+	0.3%	Flor (1956b)
	X ray	n + n	+	1.5%	Flor (1958) } Schwinghamer (1959)
	fast neutrons	n + n	+	2.0%	
Puccinia Coronata–Avena	U.V.	n + n	+		Griffiths & Carr (1961)
P. graminis–Triticum	X ray	n + n	+	1 in 1,100	Rowell et al. (1963)
	EMS	n + n	+	0.02%	Luig (1967)
P. recondita–Triticum	?	n + n	+	?	Stubbs (1968)
Cladosporium–Lycopersicon	X ray	n	+	1 in 59,000[a]	Day (1957)
Ustilago–Zea	U.V.	2n	+	1 in 2.5 × 10^5	Day et al. (1971)
Phytophthora–Solanum	U.V.	?	+?	3.7%	Wilde (1961)
	U.V.	?	−	1 in 5,000[a]	McKee (1969)

[a]Maximum sensitivity determined from reconstruction experiments.

allows expression of the mutant phenotype. Fungal parasites may be treated prior to sporulation to achieve a similar period for mutant expression (Day 1957).

The most successful experiments to date have been with cultures of *Melampsora lini* dikaryons that are heterozygous for avirulence on certain monogenic resistant flax differentials. The binucleate uredospores carried a dominant gene for avirulence in one nucleus and a recessive allele for virulence in the other nucleus for each resistance gene involved. The experiments were based on the idea that mutation or deletion of the dominant allele would lead to the expression of the recessive allele and the formation of a virulent dikaryon. Spores treated with mutagens were applied to resistant plants, and virulent mutants were isolated from the lesions that appeared subsequently. Mutation rates were surprisingly high, ranging up to 2 percent for thermal neutrons in some experiments (see Table 5.1). Only two of the virulent mutants were analyzed (Flor 1960b). They had been induced by irradiating uredospores of a culture of race 1 with X rays and selecting virulent lesions on the flax variety Koto, which carries the gene *P* for resistance (Schwinghamer 1959). The *P* locus has 3 other alleles, and these are shown in Table 5.2 together with the varieties that carry them. Two other varieties carrying alleles at another locus *M* are also shown. The genotype of the parent culture of race 1 was established by selfing and is also shown in Table 5.2, where n + n indicate the two kinds of nuclei in the dikaryon. Flor (1955) had earlier concluded that virulence on Koto, Akmolinsk, Abyssinian, and Leona was determined by separate but closely linked genes. No North American cultures of *M. lini* had been found that were virulent on both Koto and Akmolinsk, so that the two genes conditioning virulence on these two varieties were regarded as being linked in repulsion. Since race 1 was he-

TABLE 5.2
Genes for resistance in flax varieties and genotype of race 1 of *Melampsora lini*. (From Flor 1971.)

Variety	Resistance gene	Race 1 n + n	
Koto	P	Ap	ap
Akmolinsk	P1	ap1	Ap1
Abyssinian	P2	Ap2	Ap2
Leona	P3	Ap3	Ap3
Dakota	M	Am	am
Victory A	M4	am4	Am4

terozygous for virulence on both varieties, the dominant alleles for avirulence (*Ap* and *Ap1*) were assumed to be in different nuclei of the dikaryon. Flor (1960b) selfed the two mutants obtaining 155 cultures from one and 43 from the other. They were tested for virulence with the results shown in Table 5.3.

All the progeny were virulent on Koto so that the avirulent allele *Ap* present in the race 1 parent culture was not transmitted and either had mutated to a virulent allele or was deleted. The segregation for virulence on Dakota and Victory A was close to that expected for the parent culture, showing that no changes affecting virulence on these two varieties had been induced. The expected segregation also confirmed that the two mutants were not contaminants, since they were heterozygous for these two unlinked markers (*Am* and *Am4*). However, 16 cultures in the pooled progenies were virulent on Abyssinian and Leona, even though race 1 was homozygous for avirulence on them. All 16 cultures were also virulent on Akmolinsk. Flor explained the data by assuming that in both mutants *Ap* and the linked genes *ap1, Ap2,* and *Ap3,* carried on the same chromosome fragment, were deleted in one of the nuclei of the rust fungus. The original mutants were avirulent on Akmolinsk, Abyssinian, and Leona because they still carried the alleles *Ap1, Ap2,* and *Ap3* in the other nucleus of the dikaryon. Selfing gave rise to the 16 dikaryons homozygous for the deletion. One would expect a total of about 49 such cultures in a progeny of 198 (25 percent), but lowered viability of deletion homozygotes could explain the deficiency.

TABLE 5.3
Segregation for virulence on 5 flax varieties of progenies of two selfed X ray-induced mutants (A and B) of race 1 of *Melampsora lini* selected for virulence on Koto. (From Flor 1960b.)

Dakota (*M*)	Victory A (*M4*)	Akmolinsk (*P1*) Abyssinian (*P2*) Leona (*P3*)	Koto (*P*)	Number of progeny with reactions shown A	B
−	−	−	+	83	24
+	−	−	+	29	7
−	+	−	+	21	6
−	−	+	+	6	2
+	+	−	+	10	2
+	−	+	+	1	2
−	+	+	+	4	0
+	+	+	+	1	0
				155	43

− = avirulent.
+ = virulent.

It is clear from Lawrence's work (page 67) that *Ap, Ap1, Ap2* and *Ap3* are either tightly linked separate genes or alleles at one locus with a structural and functional relationship comparable to multiple-allelic resistance genes in the host. Flor's account of the inheritance of virulence in *Melampsora* may have avoided the concept of multiple allelism because it appears to impose restrictions on genetic variation like those encountered by the flax breeder. The probable effects of a deletion or null condition neatly solve this problem but at a cost to the pathogen.

If *Ap, Ap1, Ap2,* and *Ap3* are multiple alleles, the "induced-resistance" hypothesis suggests that the *Ap* locus in *Melampsora* has some other function than determining avirulence, since a nonfunctional allele (perhaps *ap*) would cover all eventualities, as it seems to do if the two mutants are both deletions. The evolution of several different avirulence specificities would serve no useful purpose unless they have other functions.

The concept that specificity differences may evolve by deletion was recently applied to the *B* incompatibility factor of the saprophytic basidiomycete *Schizophyllum commune* by Stamberg and Koltin (1973). They noted that several naturally occurring *B* factors never gave rise to recombinants by crossing-over at meiosis and attributed this to their carrying overlapping deletions that were not only responsible for the specificity differences but also for failure of recombination. Unfortunately, there is still little or no indication of the mechanism whereby these factors control incompatibility (see Fincham and Day 1971).

Some earlier results of Flor's (1956b, 1958), obtained by irradiating dikaryons heterozygous for virulence on 11 different monogenic resistant flax varieties (Flor 1971), can also be explained by the deletion of genes for avirulence either singly or in linked groups. There is also the possibility that the mutagen treatment induced mitotic recombination leading to the formation of virulent homozygotes. Only further analyses like that just detailed can establish which mechanism is responsible.

Where homozygous diploids or dikaryons, or better still, avirulent haploids, are treated to recover mutants, one avoids the complication of eliminating induced mitotic recombination as an alternative mechanism. A crucial test of the gene-for-gene hypothesis will be to see if point mutations that confer virulence towards a monogenic resistant host always involve the same pathogen locus. This test has still to be made with a pathogen like *Venturia* where the nature of the change can easily be analyzed.

In *Ustilago maydis* a comparable test of pathogenic mutants isolated from nonpathogenic diploids homozygous for the *b* locus (see page 75) showed that several pathogenic mutants recovered originated from lesions in or very close to the *b* locus. Thus, although many loci may be involved in pathogenicity, where pathogenicity depends on highly specific interactions involving one locus, we would expect that mutational changes that alter the specificity would be confined to that locus.

The reader may well ask whether attempts to induce virulent mutants in parasites are dangerous, and whether the knowledge gained from them is worth the risk that mutants may escape and render important resistance valueless. Artificially induced mutations probably differ very little qualitatively from spontaneous mutations. Mutagen treatment greatly increases the frequency of mutants, and since the effective life of a resistance gene depends in part on the mutation rate of an avirulence locus in the pathogen population, it is prudent to take precautions to ensure that any mutants produced do not escape. To do otherwise would be to negate the whole principle of plant quarantine. (See also page 172.)

To date, all published studies have set out to repeat mutations to virulence that have already occurred in nature. This is why markers are required to eliminate contaminants. The risk is that the induced mutant is either more widely virulent than anticipated or potentially so, as in the case of the two mutants of *Melampsori lini*. Even this apparently safer class of mutant should be kept under strict control and in isolation.

We can anticipate interest in another class of virulent mutants. There are occasions when a breeder is faced with a choice between two or more genes for resistance for which no virulent races are known. If circumstances prevent him from using them all, he would dearly like to predict which one will last the longest in a released new variety. If each resistance gene is triggered by its interaction with the product of a specific avirulence gene in the pathogen, and if these avirulence genes mutate at different frequencies, the resistance gene corresponding to the avirulence gene with the lowest frequency of mutation to virulence is likely to last longest. Such a prediction is very uncertain because many other factors influence the success or failure of pathogen mutant genes, that is, their frequency in the population. Although such a test could be made with mutagens, there is generally no relation between induced and spontaneous mutation rates. Thus, the exercise would not only be of little or no value but could also result in the escape of a mutant that would otherwise occur very rarely in nature.

Mutation to Avirulence

It is a fairly common experience that plant pathogenic fungi and bacteria lose the ability to produce disease on susceptible hosts after a period of growth on artificial culture media. Conditions on most culture media are clearly different from host tissue. Variants that grow more rapidly in culture because they are better adapted are much more likely to be poorer pathogens than either better or unchanged pathogens. A systematic study of spontaneous or induced cultural variants that are nonpathogenic could reveal the absence or alteration of key systems or pathways needed for pathogenicity. An alternative is to examine different classes of mutants originally selected for other properties to see what effects they might have on pathogenicity.

The first approach was begun by MacNeill and Barron (1966), who screened colonies of the storage rot fungus *(Penicillium expansum)*, produced from ultraviolet-treated spores, for avirulence to apple fruits. Auxotrophic mutants were excluded by plating irradiated conidia on a minimal medium that only supported prototrophs. Stab inoculation tests of approximately 100,000 cultures revealed 7 with altered pathogenicity. Five produced a small area of rot in comparison with the parental strain and grew more slowly in culture, whereas 2 were avirulent but behaved normally in culture. Unfortunately the basis of their avirulence was not established.

The second approach is more common and arises from the fact that investigators who have induced marker mutants in plant pathogens naturally wish to know whether the mutants are still pathogenic. Some 10 or 15 years ago, there was much interest in the possibility that the presence or absence of metabolites in hosts that would or would not support the growth of auxotrophs might explain host specificity. For the most part, the effects of auxotrophic mutants are general; a particular adenine-requiring auxotroph, for example, is equally nonpathogenic on all host varieties. Garber (1960) did show that some auxotrophs of the soft rot bacterium *(Erwinia aroideae)* are differentially pathogenic on vegetable varieties in proportion to the availability of the required metabolite. Loprieno (1964) has reviewed much of the work on induced nutritional mutants in the study in plant host-pathogen relations and found that the pattern of response varies with the host-parasite system studied.

Although spontaneous auxotrophic mutants of fungi and bacteria are sometimes isolated from nature, I am aware of only one example where auxotrophy is clearly implicated in host resistance and susceptibility. Lukezic and DeVay (1964) reported that the bark canker fungus *(Rhodosticta quercina)* requires myoinositol, and that its differential parasitism of plum varieties is controlled by the presence or absence of this substance in their tissues.

Another way of recovering avirulent mutants is the isolation of mutants with enzymic defects. For example, Mann (1962) isolated ultraviolet-induced mutants of *Fusarium oxysporum* f. sp. *lycopersici* unable to produce extracellular pectic enzymes to see whether these enzymes are important in disease development. The mutants were less pathogenic than the wild type, but they still induced wilting and vascular browning in tomato plants and could later be reisolated from the host. Mann concluded that pectic enzymes play a role in fungus invasion of the roots and vascular tissue, and she noted no root rotting with the mutants. Our knowledge of the role of these and other enzymes in pathogenesis is now somewhat more detailed, but little work with mutants has been done since Mann's. Beraha and Garber (1971) selected avirulent mutants of the soft rot bacterium *Erwinia carotovora* and then looked for revertants by making use of the fact that virulent strains grew on a desoxycholate-lactose medium but avirulent cells did not. A comparison of their cultural characteristics showed that virulence in the parental wild type and the revertant was associated with high levels of pectic, cellulolytic, and phosphatidase activities, and that these were lacking in the avirulent strain. Since the avirulent mutant and the virulent revertant both lacked enzymes for liquifying gelatin and coagulating milk, these enzymes were not implicated in pathogenicity, although they were present in the virulent wild type.

So far, no attempts have been made to select mutants that are avirulent with respect to identified genes for resistance. It seems likely that this may prove to be difficult. First, there are as yet no selective methods that can be used. The fact that mutations at a number of loci can be screened for simultaneously by looking for avirulent mutants in a race pathogenic on a host carrying many different resistance genes somewhat offsets this disadvantage. Second, one might expect to find some avirulent mutants that trigger resistance genes carried by all available host genotypes. In the absence of differentially susceptible hosts, they would be of little interest. Some avirulent mutants of pathogens already recovered may be in this category. In a sense they represent retrogressive evolution. Favret (1971) pointed out that recessive alleles for susceptibility at a locus with multiple alleles for resistance may well condition resistance to some pathogen races of exotic origin. The flax variety Bison, susceptible to all North American races of flax rust, is resistant to certain races from Australia and India (Flor 1971).

The third difficulty is that the extent of the genetic change needed may be so great that the probability of its occurrence is very low indeed. Specificity rests on the interaction of the products of a gene for avirulence in the parasite with those of a gene for resistance in the host. I earlier discussed evidence that strongly suggests that virulence may be acquired by loss of a gene for avirulence through deletion. It thus follows that virulence alleles in the pathogen population are nonfunctional or altered forms of wild-type genes, and they either form no products or form products that do not interact with host receptors. Mutation from virulence to specific avirulence is analogous to reverse mutation. In fungal genetics, forward mutation is from wild type to mutant, whereas reverse mutation is from mutant to wild type or a condition close to it.

If we think in terms of induced resistance, it seems likely that specific resistance evolves before virulence, and that resistance makes use of what parasite signals it can find. On this hypothesis, the most effective host resistance genes will be those triggered by gene products of the parasite indispensable to it, so that their loss is lethal. In another context Van der Plank has called such host resistance genes "strong genes" (see page 181). A profitable area for future study could be the development of compounds that couple indispensable parasite gene products to the host plant sensor genes for resistance mechanisms. They would be analogous to, but far more subtle than, systemic fungicides.

In considering the implications of mutation to virulence, the reader will have noticed that I have considered resistance as the key induced response involved in specificity. The virulent mutants lack an inducer of resistance. What if failure to induce susceptibility is the key to avirulence? Then we must suppose that the virulent mutant now produces an inducer, and that this is a requirement imposed specifically by the corresponding resistance gene.

The two schemes are diagrammed in Figure 5.7. The only difference between them is that the avirulent pathogen genotype produces the inducer in the first scheme, whereas the virulent one does so in the second. The first scheme most readily accounts for the rather high rates of induced mutation to virulence observed in flax rust, and for this reason it is more likely correct in this case. However, it would be unwise to conclude that all resistant interactions behave in this way. Some indications that susceptibility may be an induced response were considered earlier in this chapter.

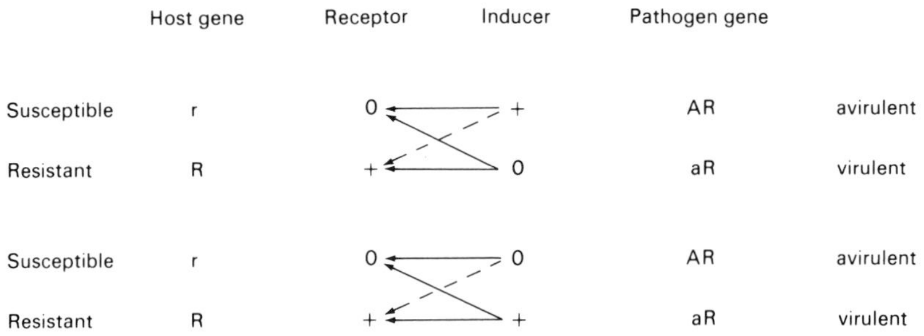

	Host gene	Receptor	Inducer	Pathogen gene	
Susceptible	r	0	+	AR	avirulent
Resistant	R	+	0	aR	virulent
Susceptible	r	0	0	AR	avirulent
Resistant	R	+	+	aR	virulent

FIGURE 5.7
Diagram of host-parasite interaction based on an induced resistant response (above), and on an induced susceptible response (below). Dashed lines indicate low infection types, and solid lines, high infection types.

It seems likely that in order to be a pathogen an organism must acquire certain properties that set it apart from obligate saprophytes. Some of these properties are likely to be interactions with the host, which predispose it to infection. Their genetic basis is likely to be complex and to have evolved over a long period of time. Avirulence could easily result from a pathogen mutation that upsets some part of this adjustment. A comparable, and as effective, move in the host to resistance would be much more difficult, since it would involve making more radical changes in metabolism than could be brought about by single gene mutations that did not at the same time greatly reduce fitness or viability. I suggest that a host plant faced with the need to evolve resistance rapidly would do so by adapting its general defense mechanism to respond to specific signals. These signals are coded by what we recognize as avirulence alleles. Thus specific resistance will generally be an induced phenomenon. On the other hand, general resistance will involve the systems that affect the balance between host and parasite metabolism. These can be more readily changed through the actions of genes of individually small effect. General resistance may thus be due to genes whose function is to block induced susceptibility rather than to induce a resistant response.

IMMUNOLOGICAL EVIDENCE

Plant pathologists have often speculated that plant host-parasite interactions might have an immunological basis comparable to that of vertebrates. Although very few workers would suggest that higher plants possess immune response mechanisms, there is, for example, evidence that specific protein-protein interactions, which have some of the attributes of antigen-antibody interactions, occur in self-incompatibility mechanisms. The pollen-style incompatibility system has the same function as the immune response, namely, the distinction of "self" from "nonself." In recent years a number of papers have appeared that show that fungal, bacterial, and viral plant pathogens and their hosts may possess common antigens as detected by the ability of experimental animals to form antibodies against them.

The more interesting of these reports (Doubly et al. 1960; DeVay et al. 1972; Wimalajeewa and DeVay 1971) suggest that where host and parasite interact to give susceptibility, their antigenic relationship is greater than if they interact to give resistance. The important question is to decide whether this is due to relatively trivial similarities, or whether it involves a recognition system and so has some bearing on specificity determination.

Recognition mechanisms in fact occur in most organisms. In a number of bacteria, there are host restriction and modification systems that mark and distinguish native DNA from invading phage DNA. Native DNA is modified after replication by methylation of a very few bases at specific sites. Methylation protects it against the cell's restriction endonuclease that is capable of recognizing and degrading any foreign DNA (Smith, Arber, and Kühnlein 1972). This mechanism is responsible for failure of a bacteriophage to grow on other cells than those of the strain on which it was produced before.

The modification of "acceptable" DNA after it has replicated appears to be strain specific. Host cell DNA and phage DNA are probably only similar to the extent to which they need to carry protective methylation.

In many fungi, recognition mechanisms, or incompatibility systems, govern genetic exchange between individuals through heterokaryon formation or mating. They may limit exchange to individuals that carry identical alleles at certain loci such as heterokaryon compatibility in *Neurospora* or sexual compatibility in *Podospora,* both Ascomycetes. Or they may limit it to individuals that carry nonidentical alleles such as sexual compatibility in *Schizophyllum* or *Coprinus,* two higher Basidiomycetes.

The application of immunological techniques to the recognition systems in fungi has not been very rewarding. Although new antigens were recovered after compatible interactions in *Schizophyllum* (Raper and Esser 1961), no differentiation of the incompatibility factors in different homokaryons was demonstrated. The new antigens were later shown to result from the production of a number of proteins identified by the altered patterns produced by acrylamide gel electrophoresis of extracts (Wang and Raper 1970). In the higher plants *Oenothera organensis* and *Petunia hybrida,* Lewis (1952) and Linskens (1960) showed with protein from disrupted pollen grains that the self-incompatibility reaction between pollen and style is correlated with specific proteins. For example, 50 percent of single pollen grains from a diploid *Oenothera organensis* plant carrying the alleles S_3S_6 produced circular areas of precipitate in thin layers of agar containing S_6 pollen antiserum (Lewis, Burrage and Walls 1967). In *Brassica oleracea,* specific antigens and characteristic protein bands were found in homogenates of stigmas from flowers of different genotypes (Nasrallah, Barber, and Wallace 1969).

In the host-parasite systems explored so far, the number of antigens that can be clearly distinguished in Ouchterlony gel-diffusion plates is small, usually 5 or 6, and rarely many more. To reach firm conclusions about similarities between host and parasite from crude extracts that probably contain hundreds if not thousands of different potentially antigenic proteins seems unwarranted unless the specific antigens are a major protein component. Wimalajeewa and DeVay (1971), working with extracts of

Ustilago maydis cells, showed that an antigenic component common to the corn host lay in the fraction containing ribosomes.

A potential source of confusion arises from the fact that most plant tissue is internally and externally contaminated by bacteria that are sampled when extracts are prepared for use as antigens. Some claims that bacterial pathogens have antigens in common with their hosts could well result from this cause. Perhaps host tissue culture can be used to avoid this hazard.

There is good evidence to suggest that during evolution, parasitic worms have become more similar antigenically to their sheep and cattle hosts (Damian 1964; Dineen 1963), but this is hardly surprising, since circulating antibodies produced by the host's immune response may be expected to play an important role in defense against infection.

For the time being the identification, by serological methods, of substances, be they proteins or other materials, that cause specificity in plant-parasite interactions remains an interesting possibility. Perhaps the most promising use of the method would be in the comparison of pathogen mutants selected for their specific effects on virulence in relation to a particular gene for resistance.

Host-Parasite Nucleic Acid Exchange

In recent years there has been a rapidly quickening interest in the evidence to show that higher plant nuclei may take up DNA of microbial origin. The possibility of transformation of normal tissue of a variety of different host plants into tumor cells by DNA from the crown gall bacterium *Agrobacterium tumefaciens* is a case in point. Bacteria-free tumor tissue of *Nicotiana tabacum* synthesizes auxin, cytokinin, and bacteria-specific RNA and antigens, none of which is formed by normal tissue. However, there is some evidence that auxin and cytokinin independence can be acquired spontaneously in normal tobacco tissue cultures. The many changes associated with the transformation are claimed to be the result of incorporation of bacterial DNA into the host nuclei rather than of the persistent derepression of part of the normal cell genome (Srivastava 1971; Srivastava and Chadha 1970). So far tumor induction has not been successfully carried out with DNA from *A. tumefaciens*. Stroun et al. (1971) suggested that the bacterial DNA carries the tumor-inducing information either as an episome or as a fragment in a temperate phage, carried by virulent strains of the bacterium. A third possibility is that the information is part of the phage DNA. Recently Ledoux et al. (1971) presented evidence that germinating seeds of *Arabidopsis thaliana* take up and linearly integrate into their own DNA tritiated DNA extracted from *Escherichia coli* or *Streptomyces coelicolor,* and that this foreign DNA can be traced to F_1 progeny. In these studies the foreign DNA was detected in cesium chloride density gradients, appearing as a heavier band than the *Arabidopsis* DNA. No genetic effects were evident, nor is it yet known whether the foreign DNA is replicated. These results suggest that transformation of tobacco by DNA from *A. tumefaciens* ought to be possible, and if viable bacteria are excluded, it should provide a critical test for the transformation hypothesis.

More recently Doy et al. (1973) have obtained evidence for the transfer of genes coding for β-galactosidase, and an amber suppressor (specifies insertion of tyrosine at amber (UAG) nonsense codons) from *Escherichia coli* to haploid tomato cell lines. The transducing phages λ and $\phi80$ were used as vectors. Although transcription, translation, and function of the bacterial genes were detected, no claim for inheritance could be made in the absence of genetic tests for transmission. The authors called the phenomenon "transgenosis."

Among bacteria pathogenic on animals, the acquisition of virulence by transformation has been known since 1928 when Griffith described it in *Pneumococcus.* The transfer of virulence from a pathogenic to a nonpathogenic isolate of *Agrobacterium* was reported by Kerr (1971), whose experiments included several convincing controls. Kerr produced galls on tomato plants by inoculating them with *A. tumefaciens* (At). The surfaces of the galls were later contaminated with a suspension of the saprophyte *A. radiobacter* (Ar). The galls were macerated after varying periods, following contamination, and plated on a selective medium that only allowed Ar cells to grow. Provided maceration was not carried out until 14 or more days after contamination, colonies selected at random and tested on tomato plants showed that up to 75 percent had acquired virulence. The virulent cells were phenotypically like Ar in the following respect: they produced 3-ketolactose from lactose; could not use erythritol, citrate, or malonate; did not require biotin; gave an alkaline reaction with litmus milk; were not inhibited by sodium selenite; gave positive reactions in oxidase, ferric ammonium citrate, and glycerophosphate tests; and carried antibiotic resistance markers. Not all Ar cultures could function as recipients; some failed completely to become virulent in extensive tests. No virulent Ar cells were recovered from *in vitro* mixtures of Ar and At cells, even under conditions that were shown to promote star formation, a form of sexual conjugation observed in *Pseudomonas* and *Rhizobium* (Heumann 1968). Kerr ruled out transformation as the mechanism of transfer on the grounds that its frequency (75 percent) was too high. Under ideal conditions the frequency of transformation never exceeds 10 percent.

However, a low rate of transformation, followed by selective multiplication of transformed cells on the host during the minimum intervening period of from 14 to 30 days, could explain his results. Previous attempts to correlate transfer with lysogeny were not successful either. The failure to obtain transfer *in vitro* could mean that the culture acquires virulence not directly from cells but from the gall tissue itself. To prove this it may be necessary to show *in vitro* that other markers are transferred by transformation or transduction but that virulence is not. It seems likely that among bacterial plant pathogens, transduction may well play an important part in facilitating genetic exchange. Evidence that *E. coli* phage DNA may be transcribed by tobacco cells (Sander 1967) also suggests that exchange between bacteria may occur with the host plant as an intermediary. Anderson (1970) has suggested that the virus mediated transfer of fragments of genome from one host to another may have considerable evolutionary significance among the many plants and animals that are virus hosts today. The discovery of RNA-dependent DNA polymerases (Baltimore 1970; Temin and Mizutani 1970) in RNA tumor viruses also suggests ways plant viruses might affect their hosts by genetic means.

6

PLANTS, PARASITES, AND PESTICIDES

The application of pesticides to control plant parasites has a long history. Insects and other animals were easily recognized as a cause of crop destruction, and so the first pesticides were insecticides and poisons such as sulfur, arsenic, and mercury. It was not until after microorganisms were recognized as a cause of plant diseases in the middle of the nineteenth century that fungicides began to come into their own. The first effective commercial fungicide, a mixture of copper sulphate and lime, was originally directed against man himself. It was applied to grape vines to discourage pilfering by passers-by. In 1882 Millardet noted that the plants so treated were green and healthy and unaffected by mildew. Bordeaux mixture, as it was called, proved very effective in trials carried out in 1885.

In plant pathology the development of fungicides has paralleled the development of resistance through breeding, and the two pursuits were carried on, for the most part, entirely independently of each other. Much the same thing happened in entomology except that insecticides became much more widely used tools for controlling insects and other arthropods than resistance, in spite of the dramatic and very important early example of *Phylloxera* resistance in grapes. The success in controlling the Colorado potato beetle by simple poisons, such as the arsenicals Paris green and London purple, in the mid- to late 1870's, no doubt led to the increasing use of this method of control.

Herbicides constitute another group of pesticides of more recent origin. They are used principally for controlling weeds and sometimes for controlling higher plant parasites. Although weeds are not parasites, they can be controlled by parasites; hence they have a place in this book.

Most pesticides are applied as sprays or powders directly to crop plants. They exploit differences between the crop plant and the parasite or target organism, so that they do not harm the one but kill, inhibit, or repel the other. The differences on which their selectivity depends may be morphological, anatomical, physiological, or even behavioral.

Bordeaux mixture worked because fungi are extremely sensitive to copper ions in solution in very low concentrations, which are often harmless to higher plants and animals. The copper was applied as a basic copper sulfate, which is almost insoluble in water and thus was not washed off by rain, and the concentration of copper ions was not toxic enough to harm vines or potato foliage. However, when used on apples, it did damage the leaves of certain varieties. Clear genetic differences exist among crop plants in their tolerance of pesticides. As we shall see, such differences later provided the basis for selective herbicides.

A phase of rapid expansion and development followed World War II when the dithiocarbamate fungicides and DDT and other chlorinated hydrocarbon insecticides were discovered and put to use. The insecticides were effective against a broad spectrum of insects, but they were also persistent. These two important features also meant, however, that they killed beneficial organisms as well as the target organisms, and that they remained to pollute the environment. Insect resistance to these compounds was an even more obvious flaw in these insecticides. It was met by applying heavier doses of chemicals more frequently, and when that did not work, by switching to other compounds. Large-scale use, coupled with the effects of accidents and carelessness, heightened public awareness of the risk to the environment.

More recently we have become concerned about the role of pesticides as mutagens towards the many organisms that encounter them. Their potential effects on man and other animals are outside the scope of this book but were recently considered by Epstein and Legator (1971) and Durham and Williams (1972).

Legislation to restrict the use of some pesticides and the banning altogether of others has stimulated research into biological controls.

Breeding for resistance, or tolerance, to insect attack is increasing. But other controls that use imported secondary parasites, population sterilization, and new trapping and luring techniques are also receiving much more attention.

The pros and cons of pesticide use are still being hotly debated. Informed opinion holds that in the future some pesticide use will be unavoidable if farmers are to produce wholesome food inexpensively. Much effort is now directed to developing highly specific pesticides that rapidly break down and are compatible with biological controls. Breeding for resistance has profound genetical consequences for host-parasite interaction. This chapter explores the genetical consequences of pesticides and biological controls.

Insecticide Resistance

Resistance to pesticides, although known earlier, appeared most dramatically in insects in response to the widespread use of chlorinated hydrocarbon insecticides such as DDT beginning in the 1940's. Since that time more than 200 different examples of resistance to insecticides have been documented. Many of these involve insects, such as house flies or mosquitoes, that are not plant parasites.

INSECT PESTS OF COTTON

The genetic consequences of large-scale insecticide applications over a period of years are well illustrated by the history of controlling insect pests of cotton in the Lower Rio Grande Valley during the last 25 years. The following account is drawn from Adkisson's (1969) summary.

After the second World War, the introduction of chlorinated hydrocarbon insecticides had a great impact on cotton production. These materials were not only highly toxic to pests when first applied, but were persistent and controlled insects that migrated from untreated areas. Cotton producers began to demand, and obtain for the first time, almost complete control of damaging pests. The consequent increases in yield and profit were spectacular. Toxaphene, BHC, dieldrin, and endrin controlled the boll weevil *(Anthonomus grandis)* but killed predators and parasites that had previously prevented the bollworm *(Heliothis zeae)* and the tobacco budworm *(H. virescens)* from becoming major pests. DDT was therefore applied with the other insecticides to replace these natural controls. All worked well for some 15 years until the boll weevil became resistant to chlorinated hydrocarbons. An organophosphorus insecticide (methyl parathion) was a satisfactory alternative, but at the dosages that controlled the boll weevil, DDT was still needed to control the bollworm and the tobacco budworm. Within three years, by 1962, these two insects had become resistant to DDT, to other chlorinated hydrocarbons, and to carbamate insecticides. They had replaced the boll weevil as the major pest.The bollworm and the tobacco budworm were brought under control again by increasing the dosage of methyl parathion from the .5 lb/acre used for weevil control to 1.0–2.0 lb/acre. Although this maintained high yields, profits fell because of the increased cost of insecticide. In 1968 the tobacco budworm developed resistance to methyl parathion. In an attempt to maintain control, farmers increased the dosages of methyl parathion. Many treated their fields 15 to 18 times and still suffered great losses in yields. Laboratory experiments showed that even large increases in dosage of methyl parathion did not kill more than 80 to 90 percent of the resistant population. As few as 2,500 budworm larvae per acre cause economic damage. In 1968 untreated population densities as high as 70,-000 per acre were found and even a 90 percent level of control, to reduce the density to 7,000 per acre, could not prevent yield losses. As Adkisson (1969) said, "The problem has come full circle. A secondary pest has achieved more importance than a primary pest and the arsenal of insecticides has almost been exhausted."

The biological unsoundness of such control measures is now widely recognized (Georghiou 1972). The 15 trouble-free years of high cotton productivity must now be paid for by searching for integrated controls where some pests are controlled by natural predators and parasites, others by breeding for host resistance, and others by insecticides which do not destroy beneficial insects. With such integrated methods, control will not always be perfect, and farmers and consumers may have to tolerate some damage, some yield reduction, and lower quality standards than they now accept.

NATURE AND GENETICS OF RESISTANCE

The range of insecticides that have selected resistant forms of crop pests in general is very large. It includes arsenate, hydrogen cyanide, nicotine sulfate, pyrethroids, dieldrin, DDT, BHC, cyclodienes, organophosphates, carbamates, and organotins. Some forms are resistant to several different insecticides by virtue of the same gene for resistance (cross-resistant). Many insect species are naturally resistant to specific insecticides.

The genetic basis of insecticide resistance has been established in a variety of examples, particularly in the house fly and several species of mosquito. Although the early work often indicated that resistance was polygenic, most examples in more recent studies show that resistance is controlled by one or a few genes. Studies of the development of insecticide resistance show that the first low levels of resistance following exposure are associated with polygenically determined vigor tolerance. Large increases in resistance generally follow after varying numbers of generations with little change. The higher level resistance is oligogenic. The biochemical basis of resistance is known in a number of these examples. Two mechanisms are common: the development of detoxication enzymes and an alteration of metabolism to avoid the effects of the block imposed by the insecticide (Oppenoorth 1965).

When resistance to an insecticide has built up in an insect population to such an extent that the insecticide is no longer used, one might hope that the relaxation of selection pressure would be followed by the eventual disappearance of the genes for resistance. There is much evidence to show that this usually does not happen even after many generations, and that the genes for resistance persist in the population and rapidly rise to a high frequency when the pesticide is reintroduced (Keiding 1967).

The evidence reviewed in Chapter 5 shows that biochemical versatility in response to insecticides is not unexpected. Many plant-eating insects have developed similar methods of avoiding the effects of toxic compounds produced by their food plants. Insects were evidently well equipped to meet the challenge of insecticides. The extent of their adaptability was recently documented by Georghiou (1972) in a review on the evolution of pesticide resistance.

Other Methods of Insect Control

The intractable problem of resistance to conventional insecticides and concern over their contribution to pollution greatly stimulated research on a variety of other methods of controlling phytophagous insects. We have already discussed breeding for insect resistance in Chapters 2 and 4. These other methods are briefly summarized in Table 6.1, which includes some examples and references to review articles. Examples of insect vectors of human or animal diseases or animal parasites are not included, although some methods not yet applied to phytophagous insects are shown.

BIOLOGICAL CONTROL

The so-called biological controls introduce a third organism into the host-parasite association that is detrimental to the primary parasite. Macrobial agents include insect parasites or predators and nematodes, and microbial agents include viruses, mycoplasms, bacteria, fungi, and protozoa. The genetic relationship between a phytophagous insect and its

TABLE 6.1

Control methods for phytophagous insects other than conventional insecticides or breeding for resistance.

Method	Examples of target insects	Reference
Predators & parasites		
insects	*Trialeurodes* (white-fly)	McClanahan (1970)
nematodes	*Popillia* (Japanese beetle)	Poinar (1971)
Microbial pesticides	—	Burges & Hussey (1971)
viruses	*Heliothis, Diprion* (larch sawfly)	Holcomb (1970)
	Porthetria (gypsy moth)	
bacteria	*Malacosoma* (tent caterpillar)	Heimpel (1967)
	Porthetria, Heliothis (tobacco budworm)	Doane (1971)
	Popillia	Falcon (1971)
		Stairs (1972)
fungi	*Malacosoma*	Stairs (1972)
		Madelin (1966)
protozoa	*Choristoneura* (spruce budworm)	Stairs (1972)
	Ostrinia (European corn borer)	McLaughlin (1971)
Sterilization		
radiation	*Laspeyresia* (Codlingmoth)	Proverbs (1971)
chemical	*Trichoplusia* (cabbage looper)	
	Anastrepha (fruitfly)	
	Hylemya (onion maggot)	Ascher (1970)
	Anthonomus (boll weevil)	
	Pectinophora (bollworm)	
genetic	—	Smith (1971)
		Whitten (1971)
hormones (arrested development)	—	Robbins (1972)
Induced resistance		
repellents	—	Beroza (1972)
antifeedants	*Aphis, Heliothis*	Ascher (1970)
synergistic compounds	—	Dyte (1967)
Behavioral controls		
traps based on response to light and other radiation	*Manduca* (hornworm)	Jones & Thurston (1970)
pheromones	*Argyrotaenia* (leaf roller)	Roelofs et al. (1970)
food lures and feeding stimulants	*Dacus* (oriental fruitfly)	Beroza (1972)
attractants		

parasites will be very similar to the relationship between the plant host and the insect parasite. Although no examples have been examined in detail, such indications as laboratory-induced resistance (see below) bear this out.

The genetic uniformity of crop plant populations will even be advantageous for biological control, since the insect pests will also tend to be genetically uniform and, hence, more vulnerable to parasites.

GENETIC CONTROL

A second major group of controls attempts to reduce the population density by releasing males that have been sterilized either by radiation or chemically. Females that mate with sterile males lay eggs that do not develop. The mass release of γ-radiation-sterilized males is extremely effective in controlling the screwworm (*Cochliomyia hominovorax*), an obligate parasite of livestock (Knipling 1955). It has been claimed that 100 or more of the major phytophagous insects which presently account for the bulk of our insecticide use could be controlled by this means. However, when applied to other insects, radiation sterilization sometimes reduced mating competitiveness. Field trials failed or were only successful if ratios of sterilized to normal males were as high as 100:1. Chemical sterilization has the same defect.

An alternative is to sterilize, or manipulate, natural insect populations by genetic means without recourse to either radiation or chemicals. A number of intriguing schemes have been suggested, but few have been put to the test. Most of the schemes are designed to introduce conditionally lethal genes. These genes only exert their effects under special circumstances. Genes for pesticide susceptibility or cold sensitivity are conditional lethals. There are several other classes. Nondiapausing mutants introduced to areas where diapause is required for overwintering from areas where it is not required might work provided diapausing control is recessive. Dominant nonfeeding mutants with either defective mouthparts or altered behavioral patterns, so that they would not seek out their natural food source, could be very useful providing they can be reared on artificial media or noneconomic hosts. Mutants of this kind might be selected by putting eggs from mutagen-treated parents on a synthetic medium with a nearby lure of the normal food, such as apples, treated with a lethal poison. Larvae remaining on the artificial diet would be tested further to screen the individuals that escaped selection from the mutants (Smith 1971).

One approach is to make use of meiotic drive, a phenomenon in which one chromosome tends to predominate in a population because the frequency of its homologue among gametes is drastically reduced. By this means conditionally lethal genes carried by such a chromosome could be introduced into the population. Segregation distorter (*Sd*) in Drosophila is one of the best known examples of meiotic drive. In males heterozygous for *Sd*, two of every four meiotic products, although motile, are nonfunctional. *Sd* and associated genes are in some way preferentially included in the functional gametes and thus occur in very high frequencies. Homozygosity for *Sd*, together with certain other genes that enhance its effects, produces sterility (Hartl 1969). Each homologue appears to suppress the other giving the effect of a recessive lethal. The lethality of homozygosity for genes controlling meiotic drive is of course a useful end in itself.

When meiotic drive is not available, other methods must be used to introduce conditional lethals. Whitten (1971) has suggested that sets of strains homozygous for different induced translocations could be introduced in sufficiently large numbers to swamp local populations. If all translocation heterozygotes are sterile, then within several generations the population will become homozygous for one of the translocations. This could be used as a method of introducing cold-sensitive mutants, which would not overwinter, or of replacing genes for resistance to pesticides with genes for susceptibility. These "useful" genes would be carried by all the introduced translocation homozygotes. Outmoded pesticides could be useful again, and the cycle could be repeated as often as necessary once the initial set of inversion stocks had been set up.

Another substitute for meiotic drive is to introduce a dominant conditional lethal with a linked gene for resistance to an insecticide normally used in the control program (Wehrhan and Klassen 1971). This has the advantage that relatively small-scale releases are likely to be effective.

The isolation, identification, and use of inversions and translocations; genes controlling meiotic drive; and conditional lethals and other linked genes for release in an insect population are techniques that call for a detailed knowledge of the genetics of the insect. This is so far available in very few phytophagous insects.

The genetic methods of population control, including sterilization by radiation and chemicals, work best when the population density of the target organism is low. The optimum ratios of introduced to wild insects can then be much more easily attained. Preliminary insecticide treatments, or trapping, can be used to reduce population levels. On the other hand, the parasite system of control generally works best when the target organism has a high population density.

The application of hormones that will arrest insects at early juvenile stages of metamorphosis can also be expected to seriously impair survival. Proponents of this method suggest that synthetic hormones, or their analogues, could be both inexpensive and highly selective, having their effect only on the target organism and even then only at certain stages of its development.

INDUCED RESISTANCE

Methods of inducing resistance or making plants repellent suffer from the hazard of applying chemicals which may have undesirable side effects. I will discuss these materials under fungicides.

BEHAVIORAL CONTROL

The mobility and behavioral responses of insects have stimulated many attempts to lure or confuse insects that are attempting to locate host plants for feeding or oviposition or to find mates for reproduction. For example, cotton plants in which the normal bracts enclosing the boll are greatly reduced (frego bract) are much less subject to boll weevil damage because the female does not readily oviposit when she is exposed and not sheltered by normal bracts (Anonymous 1971). Unfortunately the bolls of frego plants are more heavily attacked by other insect pests.

INTEGRATED CONTROL

The variety of methods of dealing with insect parasites available now and in the future make it imperative to integrate controls. Even where protection depends entirely on conventional insecticides, their application should be based on population counts and not on the calendar.

Spraying for insurance is not only inefficient and wasteful but biologically unsound, and it is partly responsible for some of our problems of insecticide resistance and pollution. Integration will mean the increasing use of highly specific pesticides that will neither harm nontarget organisms nor prejudice the use of other insects as secondary parasites. Some attempts at selecting insecticide-resistant forms of secondary parasites have been made as an alternative means to this end, but so far with little success (Anonymous 1972).

RESISTANCE

Which of these methods has already, or is likely to, run into the problem of resistance? The answer must be, any that are rendered ineffective by a genetical change that does not impair the fitness of the target insect.

Among macrobial control agents, nearly all successful uses of predators or parasites, such as the control of citrus cottony-cushion scale *(Icerya purchasi)* by the imported ladybug *(Rodolia cardinalis)*, have so far not encountered resistance. Where other attempts have failed, there is the possibility that this was due to resistance in the pest population, but this is unlikely. Failure is most frequently due to the many other factors controlling the effectiveness of the predator, such as its adaptation to the environment where it is to be established, its rate of reproduction, and so on (Remington 1968). Holcomb (1970) has commented that of more than 700 different insect predators or parasites introduced, less than a quarter have become established. One instance where control failure was probably due to resistance involves the larch sawfly (*Pristiphora erichsonii*) in Canada. Although satisfactory levels of infection with the parasite *Mesoleius tenthredinis* were obtained in British Columbia, in Manitoba and Saskatchewan 100 percent of the parasite eggs were encapsulated in the sawfly host larvae and were unable to develop further (Muldrew 1953).

So far no resistance has appeared in the field to microbial control agents, probably because most of them have been in use for a comparatively short time. Several laboratory studies have shown that resistant mutants can be produced experimentally, such as cultures of cabbage worm (*Pieris brassicae*) resistant to granulosis virus disease and housefly (*Musca domestica*) resistant to the exotoxin of *Bacillus thuringiensis* (Burges, in Burges and Hussey 1971). Almost certainly as microbial agents are used on an increasingly large scale, especially if they are persistent and so exert a steady selection pressure, resistant forms will appear.

The large-scale introduction of myxomatosis virus in Australia to control the rabbit was followed within seven years by the development of resistance to the virus. At the same time, less virulent forms of the virus developed which provided adequate opportunity for the mosquito vector to become infected (Fenner 1965). This example points to an important feature of biological control agents: their ability to change.

Experience with insecticides would suggest that resistance to chemical sterilants could be expected, and it has indeed been found in houseflies and mosquitoes (Proverbs 1969). With regard to the other genetic methods, there could be a danger that the introduction of a foreign genotype could result in the rapid development of isolating mechanisms and lead to speciation. Necessarily small-scale laboratory experiments could fail to reveal niche preferences or other features in introduced stocks that might reinforce these tendencies. The use of synthetic sex attractants as lures would seem unlikely to generate resistant mutants not attracted to traps or poison stations, since these forms would have difficulty in finding mates. However, there are indications that sex attractants are not as efficient as was hoped. For example, although males of the gypsy moth (*Porthetria dispar*) respond to "disparlure," a synthetic sex attractant, there appear to be other signals from the female that make her more attractive than the lure.

Fungicide Resistance

Until recently it was true to say that fungicide resistance among plant pathogens was strikingly less frequent than insecticide resistance among phytophagous insects. Table 6.2 lists 7 examples prior to 1969, all that I could find, and 7 examples published since then of fungicide resistance, reported from the field, that were either shown to reduce the efficiency of control or were suspected of doing so. The genetic basis of resistance has not been established in any of these examples.

The most widely accepted explanation for the rarity of fungicide resistance prior to 1969 was the nonspecific nature of many of the most widely used fungicides. This means that generally no single genetic change could result in high levels of resistance. For example, the diethyldithiocarbamates inhibit 20 or more different enzymes which, presumably, is the basis of their fungitoxicity. It is extremely unlikely that mutants would develop in which the 20 or more blocked activities are restored either by variant enzymes, whose activities are unaffected, or by alternative metabolic pathways to relieve the blocks, or a combination of the two. In theory, resistance could result if fungus cells became impermeable to or

TABLE 6.2
Examples of fungicide resistant plant pathogens isolated from the field.

Organism	Fungicide	Reference
Prior to 1969		
Physalospora obtusa	Bordeaux mixture	Taylor (1953)
Penicillium digitatum	diphenyl, sodium	Harding (1959)
P. italicum (post-harvest citrus rot)	orthophenylphenate	Duran & Norman (1961)
Sclerotium rolfsii	pentachloronitrobenzene	Georgopoulos (1964)
Tilletia caries *T. foetida*	hexachlorobenzene	Kuiper (1965)
Rhizopus arrhizus	botran	Weber & Ogawa (1965)
Sclerotinia homoeocarpa	cadmium succinate	Cole et al. (1968)
Pyrenophora avenae	phenyl mercuric acetate	Malone (1968) Greenaway (1971) Sheridan (1971)
1969 and after		
Sclerotium cepivorum	2,6-dichloro-4-nitroaniline (botran)	Locke (1969)
Venturia inaequalis	n-dodecyl-guanidine acetate (dodine)	Szkolnik & Gilpatrick (1969)
Sphaerotheca fuliginea	benomyl	Schroeder & Providenti (1969)
Botrytis cinerea	benomyl	Bollen & Scholten (1971)
Cercospora beticola	benomyl	Georgopoulos (pers. commun.)
Sphaerotheca fuliginia	dimethirimol	Dekker (1971)

detoxified these materials, but neither kinds of resistance have developed in the field, although detoxification is known to be a reason why some pathogens are not effectively controlled by certain phytoalexins (see page 121).

Georgopoulos and Zaracovitis (1967) pointed out that the fungicides which most frequently selected resistant mutants had in common an aromatic hydrocarbon nucleus (Figure 6.1). Cross-resistance is very common

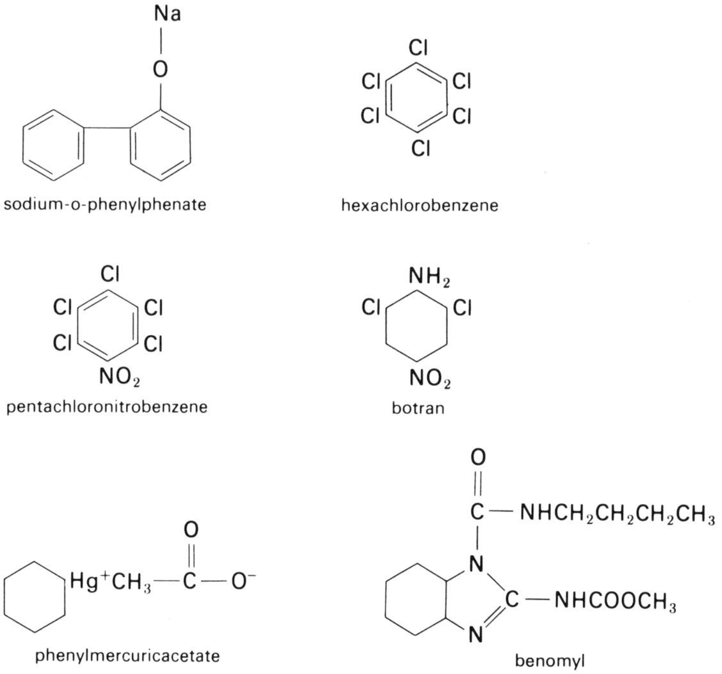

FIGURE 6.1
Fungicides with aromatic hydrocarbon nuclei known to have selected resistant mutants.

in this group. For example, mutants of *Nectria haematococca* var. *cucurbitae* (formerly known as *Hypomyces solani* f. *cucurbitae*) selected for resistance to one of the five compounds diphenyl, sodium orthophenylphenate, hexachlorobenzene, botran, or PCNB are also resistant to the other four, and to some other compounds, including naphthalene and acenaphthene. The nature of action of aromatic hydrocarbon fungicides is unknown. Cross resistance could depend on impermeability to molecules containing aromatic hydrocarbons, but to my knowledge, this has not been tested.

NATURE AND GENETICS OF RESISTANCE

Few genetic studies of fungicide resistance have been carried out in plant pathogenic fungi and only on mutants of laboratory origin. Examples in *N. haematococca* and *U. maydis* are noted in Table 6.3. These, and other examples in nonpathogenic fungi (reviewed in Georgopoulos and Zaracovitis 1967; see also Hastie and Georgopoulos 1971), show that resistance is commonly determined by single genes and that several unlinked loci may control resistance to the same fungicide. Two recent examples illustrate the biochemical mechanisms of resistance.

In *Ustilago maydis* the oxathiin derivative carboxin (full name given in Table 6.2) was used to select resistant mutants induced by UV irradiation of haploid cells (Georgopoulos and Sisler 1970). Two kinds of mutants, determined by the unlinked genes *oxr* and *ants* (antimycin-A sensitive) were recovered. The first, with a high level of resistance, possessed a greater succinate dehydrogenase (SDH) activity in the presence of carboxin. The high SDH activity was also shown by extracts from mutants, and Georgopoulos (unpublished) concluded that in the mutant the SDH activity was in some way protected, and that resistance was not due to a decreased permeability to carboxin. Since SDH is carried in the mitochondria, tests were made to see whether wild-type and mutant mitochondria might be distinguishable by other means. Mitochondrial preparations of mutant cells were readily distinguished from those of the wild type by their lower NADH-ferricyanide reductase activity, although the *in vivo* rates were the same.

The *ants* mutant was only slightly more resistant to carboxin than the

TABLE 6.3
Fungicide resistant forms of plant pathogens recovered in laboratory studies.

Organism	Fungicide	Treatment	Genetic basis	Reference
Botrytis cinerea	various		?	Parry & Wood (1959a, b)
	botran		?	Webster et al. (1970)
B. allii	chlorinated nitro-benzenes		?	Priest & Wood (1961)
Nectria haematococca var cucurbitae	PCNB & TCNB	—	3 loci	Georgopoulos (1963)
	dodine	U.V.	4 loci	Kappas & Georgopoulos (1970, 1971)
Rhizopus stolonifer	2, 6-dichloro-4-nitroaniline	γ radiation —	?	Webster et al. (1968)
Cladosporium cucumerinum	6-azauracil	U.V.	?	Dekker (1969)
Ustilago maydis	carboxin (5, 6-dihydro-2-methyl-1, 4-oxathiin-3-carboxanilide)	U.V.	2 loci	Georgopoulos & Sisler (1970)

wild type, but was sensitive to the antibiotic antimycin-A which blocks electron transport involving cytochromes *b* and *c*. The wild type is insensitive to antimycin-A. Respiration of the mutant was strongly inhibited by azide and cyanide at concentrations that stimulated respiration of the wild type. Evidently an alternate electron transport path is missing in *ants,* although why it should show a low level of resistance to carboxin is not clear.

The second example, illustrating another mechanism of resistance, is a UV-induced mutant of *Cladosporium cucumerinum,* resistant to the experimental systemic fungicide 6-azauracil (AzU) described by Dekker (1969). In wild-type cells, AzU is converted to 6-azauridine-5′-phosphate (AzUMP) by the action of two enzymes: uridine phosphorylase, which produces the intermediate 6-azauridine (AzUR), and uridine kinase, which converts this to AzUMP. The end product AzUMP is toxic and inhibits the decarboxylating enzyme, which converts orotidine-5′-phosphate to uridine-5′-phosphate. The resistant mutant could not convert uracil into uridine; it was sensitive to AzUR and AzUMP, and so evidently lacked uridine phosphorylase, the first enzyme in lethal synthesis (Figure 6.2).

An example of fungicide detoxification was described by Bartz and Mitchell (1970). *Fusarium solani* f.sp. *phaseoli* is some 30 times more

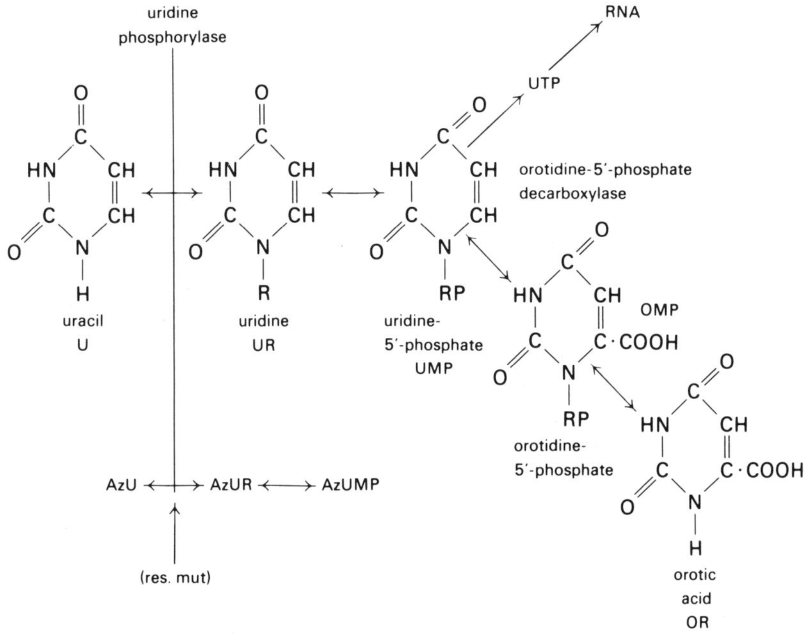

FIGURE 6.2
Pathways of uridine-5′-phosphate synthesis and position of block in mutant resistant to 6-azauracil in *Cladosporium cucumerinum. (After Dekker 1969.)*

tolerant of dodine than *Venturia inaequalis,* as judged by ED_{95} values for spore germination. These authors suggest that this may be due to the detoxification of dodine that occurs after its uptake by *Fusarium.* Spores of *F. solani* took up ^{14}C labeled dodine, from concentrations that inhibit germination, and released a ^{14}C labeled derivative that was partially characterized and had a markedly lower toxicity. Whether dodine resistance was a feature of other isolates of *F. solani* was not reported. The biochemical basis of the dodine resistance in *V. inaequalis* from the field (see Table 6.2) is still unknown.

Biological Controls

Breeding for resistance is a biological method of controlling parasites. In this case the environment represented by the potential host plant is rendered unfavorable for parasite development by genetic manipulation. Similar results can often be obtained by nongenetic means, such as deliberately introducing or encouraging the growth of organisms that are antagonistic to parasites but do not harm the host. This alternative method of control was comprehensively surveyed by Baker and Cook (1974). Although a full treatment is beyond the scope of this book, it is important to note that the two approaches may sometimes achieve their ends by surprisingly similar means. Buxton (1957) pointed out that pea varieties resistant to wilt caused by *Fusarium oxysporum pisi* produced effects on the soil adjacent to the root system (rhizosphere), promoting growth of microorganisms that were antagonistic to the pathogen. More recently Neal et al. (1970) compared two lines of spring wheat *(Triticum aestivum).* One line was susceptible to common root rot *(Cochliobolus sativus);* the other was resistant by virtue of a pair of substituted chromosomes. In other respects the two lines were nearly isogenic. Tests of rhizosphere organisms, isolated from soil particles clinging to roots of the two plant types, indicated that resistance was associated with the presence of bacteria antagonistic to *C. sativus.* Such effects could well be common in a variety of crop plants.

The deliberate introduction of antagonistic microorganisms to control plant diseases has hardly begun. Wells et al. (1972) isolated a strain of *Trichoderma harzianum* from diseased sclerotia of the soil fungus *Sclerotium rolfsii.* They showed that inoculum of *T. harzianum* applied to the soil surface around tomato transplants reduced damage from *S. rolfsii.*

Induced Resistance to Parasites

Systemic pesticides are taken up by plants at the point of physical contact (leaves, roots, or inoculated stems) and transported to other organs, or even to the infection or infestation sites, which they protect by making them toxic to invading parasites. Some systemic materials are highly specific; for example, the oxathiin fungicides are effective only against Basidiomycetes. So far, little work has been published on materials that are not toxic by themselves but either increase the toxicity of materials already present in the plant by synergistic interaction (Dyte 1967) or are transformed by the plant into toxic materials. A third category may be materials that realize the host's innate disease resistance mechanism so that a potentially resistant host invaded by a virulent physiologic race would have its resistance restored (Day 1968).

Materials in the third category could well be highly specific, and therefore, carry less risk of unwanted side effects. They would probably also be effective against the same parasite in plants not known to possess any genes for resistance. This is likely for two reasons. First, such plants will almost certainly carry resistance genes that are undetected simply because the available parasite population is uniformly virulent (see, for example, the flax variety Bison, page 147). Second, the compound would effectively replace a resistance gene by establishing a link between the parasite and the host defense mechanism, thus enabling the host to recognize the parasite as an intruder and respond accordingly. In looking for such materials, the precedent of effective oligogenic resistance in a given host-parasite interaction would be encouraging but not a requirement. It is possible that compounds with this kind of activity are known already—for example, pesticides with low *in vitro* toxicity that are effective *in vivo*.

Nonspecific restoration of resistance is also possible. Királi et al. (1972) showed that in the presence of chloramphenicol virulent races of *Phytophthora infestans* and *Puccinia graminis tritici* produced necrotic reactions on their respective hosts. Why this occurred is still unknown; perhaps the drug interfered with the induction of susceptibility.

Another form of induced resistance results from prior inoculation of the host tissue with an avirulent pathogen. Some examples that are discussed elsewhere include *Phytophthora infestans* on potato tuber tissue (page 116) and *Melampsora lini* on flax (page 135). These experiments confirm that in many cases once resistance has been induced it is effective against virulent and avirulent forms alike.

Pesticides as Host and Parasite Mutagens

The discovery that the large-scale use of persistent synthetic pesticides like DDT soon led to the appearance of resistant target organisms naturally raised the question whether the pesticides were mutagenic. Initial tests of

DDT with *Drosophila melanogaster* suggested that DDT was not a mutagen (Lüers 1953). Current opinion is that lethal doses of these materials select spontaneous variants in the population. However, the partial literature review recently carried out by Epstein and Legator (1971) shows that of some 32 pesticides tested, 31 showed activity in one or more test systems. DDT, for example, was listed as increasing the frequency of dominant lethals in rat sperm, and causing chromosome breaks and mitotic abnormalities in two different test plants. In regard to the latter, we may note that Vaarama (1947), whose work was cited, claimed that the major effects on root tip mitosis were produced by ethanol or ethanol plus DDT. Used alone, DDT caused chromosome contraction at metaphase and no chromosome breaks.

The current awareness of the mutagenic and related hazards of pesticides should rapidly fill some of the many gaps in our knowledge. Although present concerns are with effects on man at the end of the food chain, we should not overlook other interactions.

For example, benomyl, a carbamate effective in controlling many phytopathogenic fungi, was shown to induce instability in diploids of *Aspergillus* (Hastie 1970b). In this respect it is like the phenylalanine antimetabolite p-fluorophenylalanine (see page 42). Haploidization resulting from diploid instability is an important part of the parasexual cycle in fungi, so far known only in the laboratory. Benomyl could accelerate recombination, and therefore genetic change, in fungi. At the present time we cannot say whether such an effect is common or even important. We simply do not have the information. The question nevertheless should arise each time a new material is being developed for release. Our present uncertainties over such questions provide much of the impetus to breeding for resistance and developing biological controls.

Herbicides and Weed Control

SYNTHETIC HERBICIDES

Weed control specialists have come to depend heavily on synthetic chemical herbicides during the years since World War II. Many of these compounds are selectively phytotoxic. They have little or no effect on a particular crop but kill its sensitive weeds. The biochemical basis of phytotoxicity is too large a subject to concern us here but is covered in several reviews (Moreland 1967; Audus 1964). Selective phytotoxicity relies on some form of herbicide resistance in the crop plant.

Resistance may be due to failure to absorb the herbicide (because the leaf shape and surface allow little or no retention), to insensitivity, or to detoxification by conversion to an inactive derivative. The study of how chemical pesticides in general are metabolized by higher plants is a rapidly developing area of plant physiology (Casida and Lykken 1969). It would not be surprising to find that the use of selective herbicides results in the selection of resistant forms of formerly sensitive weeds. But resistant mutants seem not to have occurred. There are two reasons for this. First, most weed plants can only complete one generation a year, and compared with insects or microorganisms where resistance to toxic materials is frequently encountered, few generations have been exposed to one selective toxic agent. Second, a selective herbicide very quickly selects other weeds that are resistant. As these become dominant, they must be controlled by using other herbicides or other methods. Thus any resistant mutants selected from the original population will also be destroyed.

Intrinsic herbicide resistance is the weed specialist's biggest problem. The worst weeds from this point of view are purple nut sedge *(Cyperus rotundus)* and Johnson grass *(Sorghum halepense),* both extremely difficult to control with herbicides.

Some agronomists have advocated selecting herbicide resistant crop varieties with a view to expanding the possibilities open to them in weed control. Mutants of wheat and tomato resistant to the herbicides terbutryn and diphenamid, respectively, were recovered from plants grown from seed treated with ethyl methanesulfonate (Pinthus et al. 1972).

NATURAL HERBICIDES

It is well established that some higher plants excrete materials differentially toxic to other plants. The phenomenon is known as allelopathy, and its role in the evolution of vegetation was recently reviewed by Muller (1970). There seems to have been no conscious practical application of allelopathy for the reason that rather few crop plants, apart from the black walnut *(Juglans nigra)* and black mustard *(Brassica nigra),* are known to exhibit it. The toxic materials usually prevent germination, or seedling growth, of nearby sensitive species. Some phytotoxins have been identified; for example, the walnut toxin (juglone) is 5-hydroxy-1, 4-naphthoquinone. *Salvia leucophylla* and *Artemisia californica* produce terpenes that inhibit seedling herbs in annual grassland in the coastal valleys of Southern California. Other allelopathic shrubs produce highly toxic phenolics that are washed off the leaves and stems by rain and accumulate in the soil and leaf litter. Several important pioneer weed species are allelopathic. For example, *Sorghum halepense* produces phenolic compounds that not only inhibit seedlings but also inhibit nitrogen-fixing and nitrifying soil bacteria.

Allelopathic interactions are not uncommon. They suggest the possibility of breeding crop plants that are weed-resistant by virtue of allelopathic effects.

BIOLOGICAL WEED CONTROL

Man's experience in protecting his crops from pests and parasites is increasingly being put to use in reverse by introducing and encouraging parasites of weed species. The exotic parasites range from bacteria with surface-bound enzymes that aid them in attacking summer blooms of blue-green algae in lakes and reservoirs (Daft and Stewart 1971) to herbivorous fish for controlling aquatic weeds (Michewicz et al. 1972).

Biological control of weeds is still in its infancy compared with its antithesis, breeding for resistance. All biological control agents in current use have been selected from nature, usually from the place of origin of the weed. So far none have been produced by mutagen treatment followed by screening for activity. This could be profitable if resistance to biological agents appears, providing certain precautions are taken to ensure that a dangerous parasite capable of attacking related nonweed hosts is not released. There is also the risk of a parasite spreading to other areas where the plant in question is not a weed. For example, the persimmon is a pasture weed in some regions of the United States but is of economic value in other regions. Mutagen treatment could also be used to increase both the virulence and host specificity of a weak parasite with too broad a host range.

Biological control should be particularly effective against weeds that are relatively uniform genetically by virtue of limited outcrossing, apomixis, or vegetative reproduction. These species are less likely to produce resistant variants.

The basic problem in weed control has no close analogy with parasite control. Weeds of arable land grow in soil, generally in the spaces between crop plants. When a weed plant is destroyed, its place tends to be taken by another of a different species. The soil is a far less selective substrate than the tissue of a host plant. It can often support many different kinds of weeds. The succession of weed destruction followed by replacement with another taking place over several seasons may proceed to the point where the problem becomes intractable, with a weed like nut sedge firmly established. Biological controls could be most effective in dealing with these "climax" weed situations, which are relatively stable because previous treatments have exhausted the other possibilities. The eradication of the mission prickly pear *(Opuntia megacantha)* from Queensland by the introduction of the insects *Cactoblastis cactorum* and *Dactylopius confusus* is a good example of success where other methods had failed.

The reader interested in examples of biological weed control should consult Wilson (1969) on the use of plant pathogens, and B. E. Day et al. (1968) on the use of insects.

7

GENETICS OF EPIDEMICS

In this chapter I describe how the individual genetic components of host-parasite interaction work together at the population level, and how they encourage or prevent epidemics and epizootics. The development of an epidemic integrates all the variables that act and interact on each other. For the host one must consider plant size, maturity, the spacing, the leaf area, and the fruit load. For the pathogen one must consider spore germination, host penetration, host tissue invasion, size of lesion, reproduction, and movement to new hosts. Then one must consider the environment: light, darkness, day length, temperature, humidity, dew, rainfall, and windspeed. Nineteen variables are listed, providing 361 possible interactions going on continuously and changing over time, as the day waxes and wanes and the highs and lows pass over. Even the simplification of assuming that host and parasite populations are genetically homogeneous leaves a task so complicated that only a computer can handle it. My colleagues Waggoner and Horsfall (1969, 1972) have developed computer programs that simulate the progress of plant disease epidemics. Their programs use meteorological information to describe or predict with considerable accuracy the progress of *Alternaria solani* in a tomato crop or *Helminthosporium maydis* in a corn crop. The task of writing the programs for these two fungi revealed some interesting gaps in knowledge of the biology of both host and parasite that had to be filled by experiments.

The next step of designing computer programs that take genetic variation into account has still to be taken. What follows is a discussion of the genetic factors and their effects.

Several books, reviews, and articles deal with the technical aspects of epidemiology (Van der Plank 1963, 1968; Waggoner and Horsfall 1969; Waggoner et al. 1972), and I shall only draw on these topics as they are needed to develop my theme.

Essentials of Epidemics

The prime genetic basis of plant disease epidemics can be seen by examining two examples to see how they started and how they ran their course. The first, on oats, happened 25 years ago; the second, on corn, happened in 1970 and is still fresh in our minds. In both examples, the disease was caused by a *Helminthosporium,* a genus whose propensities were discussed in Chapter 3.

VICTORIA BLIGHT—1946

In 1942 an oat variety resistant to all races of crown rust (*Puccinia coronata*) known at that time was introduced in Iowa. The new variety, called Victoria, derived its resistance from a collection of *Avena byzantina* from Uruguay, made in 1927. Victoria was also resistant to stem rust (*P. graminis avenae*) and two smuts (*Ustilago avenae* and *U. kolleri*). Victoria and its derivative varieties Boone, Tama, and so on, were grown on a large scale, until by 1945 they made up 97 percent of the total oat acreage.

In 1946 a seedling blight appeared which in two years reached epidemic proportions. The causal fungus, *Helminthosporium victoriae,* was later found to be a minor parasite of wild grasses that was widely distributed (Scheffer and Nelson 1967). Victoria was peculiarly susceptible to *H. victoriae,* apparently because of a pleiotropic effect of the dominant gene *Pc-2* for crown rust resistance (see pages 17–19 in Chapter 2). Its cultivation on a large scale was responsible for the selection and rapid multiplication of the oat blight fungus.

In 1947, the year following the appearance of *H. victoriae,* the Iowa acreage planted with Victoria and its derivatives was reduced by over 20 percent. It would have fallen further had there been seed available of a satisfactory substitute oat variety. Within several years Victoria had been almost entirely replaced by Bond and related varieties, until these too were subject to increasingly severe attacks of crown rust and were themselves replaced by the resistant varieties Landhaufer and Santa Fe. In other parts of the United States, however, Victoria derivatives were unaffected by blight and continued to be grown for some years.

The critical points are: (1) Victoria and its derivatives came to occupy the bulk of the oat acreage so that the crop was genetically uniform for *Pc-2*, (2) the epidemic was favored by weather conditions in 1946 and 1947, and (3) Victoria blight ceased to be a problem in Iowa as soon as new resistant varieties were introduced.

SOUTHERN CORN LEAF BLIGHT—1970

This disease caused by *Helminthosporium maydis* has been a minor cause of leaf spotting in corn for many years. In 1970 it reached epidemic proportions, causing severe losses especially in the south eastern region of the United States. Plant breeding had created a corn monoculture with consequences strikingly similar to those arising from the oat monoculture based on Victoria. The chief difference was that the corn plants had uniform cytoplasm, whereas the oat plants all carried the same gene for crown rust resistance.

The first commercially feasible hybrid corn followed D. F. Jones's demonstration in 1917 that crosses between certain F_1 hybrids could produce high yields of hybrid seed of normal size. Although the double cross hybrids were high yielding, their acceptance by farmers and seedsmen was slow. By the mid-1930's, however, they began to be grown on a very large scale. In hybrid seed production, self-fertilization of the seed parent was prevented by removing the tassels or male flowers which grow at the top of the plant. This was both expensive and inconvenient, and the introduction of genetic male sterility as an alternative was an important advance. Jones and Mangelsdorf developed cytoplasmic male-sterile inbred lines that were released in 1946. In these lines male sterility is transmitted by the female parent to all the progeny. A male-sterile inbred cannot be maintained by selfing but has to be crossed with a male-fertile, but otherwise identical, inbred. Existing fertile inbreds are easily converted to the male-sterile form by backcrossing. The inbred to be converted is used as a recurrent pollen parent. For commercial hybrid seed production, two rows of the male-fertile pollen parent alternate with 4 rows of the sterile or seed parent across the field. Wind carries pollen from the tassels of the male parent to the silks of the seed parent. The sterility of the hybrid can be overcome in two ways. One is to use an inbred line homozygous for dominant pollen fertility-restoring genes as the male parent so that the hybrid is male-fertile. The other way is to add a proportion of male-fertile seed of the same hybrid made up by hand detasseling. In a crop grown from the resulting blend, sufficient pollen is produced by the randomly distributed fertile plants to fertilize all the fertile and sterile plants in the field.

Two kinds of cytoplasmic male sterility were released initially. One had been discovered by Mangelsdorf and Rogers at the Texas Agricultural Experiment Station in plants of the varieties Mexican June and Golden June. The other was discovered in corn material collected by M. T. Jenkins of the USDA and sent to Jones. The two kinds of male sterility, distinguished at that time by the fact that they came from different sources and responded to different restorer genes, were designated T (Texas) and S (sterile).

Extensive tests showed that cytoplasmic male sterility carried no penalty in reducing yield or quality or in introducing disease susceptibility. The method was rapidly adopted by the seed industry. Texas male sterility, or *Tms,* proved to be more reliable than *Sms.* Plants carrying *Sms* were sometimes partially fertile and could thus produce some nonhybrid seed. *Tms* was introduced into many inbred lines during the 1950's, and the scale on which it was used increased until, by 1969, from 70 to 90 percent of the field corn grown throughout the United States was based on *Tms* cytoplasm.

Hybrid corn made up in this way was not acceptable everywhere. In the Philippines, Mercado and Lantican (1961) reported that *Tms* hybrid corn was very susceptible to *H. maydis,* but hybrids with normal cytoplasm were not badly diseased. The *Tms* hybrids were susceptible whether or not they carried fertility-restoring genes. Tests carried out in the United States several years later failed to demonstrate a similar susceptibility due to *Tms* cytoplasm (Hooker et al. 1970). Mercado and Lantican's paper was important because it was one of the first descriptions of cytoplasmic inheritance of disease resistance, but its full significance was not realized until 9 years later. In 1969 reports appeared that *Tms* lines and hybrids in the United States corn belt were unusually susceptible to yellow leaf blight (*Phyllosticta maydis*) (Scheifele et al. 1969). A *Phyllosticta* leaf blight was first described in 1930, but it never attracted attention in the United States corn belt until 1968. It became important perhaps because *Tms* cytoplasm was used on so large a scale. One major hybrid seed producer had decided not to use *Tms* in 1970 because of its susceptibility to *Phyllosticta.* Late in 1969 an abnormally high incidence of southern leaf blight was reported on corn in several midwestern states. But in late February and early March of 1970, southern leaf blight reached epidemic proportions in Belle Glade, Florida, and from there moved steadily to Georgia and spread west around the gulf coast of Texas. Moving north in two prongs, a broad path centered on the Mississippi River to Wisconsin and southern Minnesota, and along the coastal plains of Georgia and the Carolinas to New England, until by

September southern leaf blight had been reported throughout the eastern half of the United States.

It quickly became obvious that the varieties showing severe blight symptoms were those based on *Tms* cytoplasms. Where fields were sown with blended seeds, only the *Tms* plants were heavily blighted. In heavily blighted areas, only hybrids with normal cytoplasm, produced by detasselling, gave crops that were relatively undamaged.

The epidemic was produced by race T of *H. maydis*. Until 1969 only one form had been recognized, now known as race O. Race O does not differentiate between *Tms* and normal cytoplasm. Mercado and Lantican presumably encountered race T in the Philippines. In 1968 Nelson and Kline examined 150 different isolates of *H. maydis* from a number of different geographic areas for their capacity to blight various grass hosts. The material, stored since collection as dried infected host leaves, was reexamined by Nelson et al. (1970), who found that 24 isolates, one collected as early as 1955 from North Carolina, produced symptoms characteristic of race T when inoculated to corn. The two mating types occurred in approximately equal frequencies in the sample of 24. Evidently forms like race T existed and were widespread before the epidemic of 1969–70.

We might ask why the epidemic took so long to develop in the United States if race T already occurred in North Carolina in 1955. It is true that the incidence of *Tms* cytoplasm was probably greater in 1970 than in 1955. But the use of *Tms* in the Philippines in 1961 was not extensive and yet race T appeared there. Were the weather conditions in the corn belt exceptional in 1970? There is no question that they favored southern blight, but meteorological records show they were not unusual in this respect. Although isolates capable of producing lesions on *Tms* corn were present, it seems likely that a genotype which could produce an epidemic was not. It arose later as a result of mutation or recombination or even introduction from elsewhere.

Some support for this suggestion comes from a determination of race and mating types in a sample of 71 isolates, collected in 1970 in the midwest and southeast by Leonard (1971). Of the 66 that were race T, 58 (88 percent) were mating type *A* and could thus have had a common origin. A subsequent test of isolates collected in 1971 again showed a predominance of race T (Leonard 1973). However, in the South the proportion of T isolates that were mating type A had shifted from 80 percent in 1970 to 47 percent in 1971. Leonard (1973) attributed this shift to selection for genes affecting fitness linked to the *a* mating-type factor. However, a greater participation of the perfect stage in overwintering in this area than is presently known would provide an alternative explanation.

In 1970 seed producers and breeders took several steps to prepare for 1971. Hybrids with normal cytoplasm were resistant to races O and T, and their production by detasseling was increased as much as possible. Although the seed stocks were only sufficient to supply slightly more than 20 percent of the total demand for resistant seed, the weather conditions in 1971 did not favor an epidemic and damage to *Tms* corn was minor (Ullstrup 1972).

Other cytoplasmic male-sterile lines, distinguished by the specific restorer genes which render them fertile, were tested for resistance to races O and T, and a number were released to plant breeders. At the same time, the influence of nuclear genetic background on susceptibility to race T was clarified, and the prospect developed of selecting parents that transmitted resistance that was expressed in T cytoplasm. At the time of writing the situation is fluid. Most seed producers have returned to hand or mechanical detasseling until such time as resistance, new sterile cytoplasms, or new methods (such as chemical male gametocides) persuade them to change.

An important outcome of the southern leaf blight epidemic was the realization of just how precarious the position of this major food crop had become. More than 70 percent of a crop covering some 69 million acres in the United States was cytoplasmically uniform. The same trend to genetic uniformity was found to be true of other major crops (Horsfall et al. 1972, Day 1973).

All epidemics require a high degree of genetic uniformity among their host populations. Breeding for disease resistance introduces important genetic variation to the crop, which is designed to prevent epidemics. The traditional approach employs a succession of monogenic resistant varieties where each one is replaced as it fails. This relieves genetic uniformity by creating genetic discontinuities in time. An alternative is to create discontinuities in space by deliberately introducing variation by growing mixed populations. This is discussed later in the chapter (see Multilines, pages 184–185).

Gene Frequencies—Mutation Rates

The genetics of host-parasite populations describes the relationship between frequencies of resistance genes in the host and frequencies of virulence genes in the parasite. Resistance gene frequencies in crop plants are controlled by the varieties that are grown. Crops and varieties are chosen according to their adaptation to the environment, yielding ability, quality, resistance to parasites, and other factors that determine the economic return. The kinds of parasites that occur on the crop are also determined by the environment, by the cultural methods employed, and by the varieties used.

The useful life of a race-specific resistance gene deployed against a crop parasite is determined by the frequency of the allele for virulence in the parasite population. This, in turn, is determined by the mutation rate to virulence, and by the fitness of the virulent mutant in competition with avirulent nonmutants and other organisms trying to occupy the same niche. It would obviously be useful to know whether virulent alleles are likely to arise by mutation—and if so, how frequently. I discussed part of this question on page 145 in connection with induced mutation to virulence. Slootmaker (personal communication) recently made some calculations for the powdery mildew fungus (*Erysiphe graminis hordei*) based on spontaneous mutation rates that bear on this question. He assumed that a mature sporulating lesion of the mildew produces 10^4 conidia per day, and that, with 10 percent of the leaf area infected, giving some 10^5 lesions per square meter, the daily spore production per hectare would be 10^{13}. If the rate of spontaneous mutation per locus in the mildew is comparable to the average rate of reverse mutation in *Neurospora*—about 10^{-8} (Fincham and Day 1971)—then some 10^5 conidia that are potentially virulent on a nearby newly introduced monogenic resistant barley variety are produced each day. Of course, since there are 10^8 as many avirulent conidia as there are virulent conidia, the probability that a virulent mutant conidium will infect a host at all is still extremely low. Robinson (1971) has argued that mutation rates to virulence vary among pathogens, and that very low rates could result in continued protection by a single resistance gene for many years. Practical experience does not usually enable one to discriminate between low mutation rate and low fitness of a majority of virulent mutants since the end points are likely to be the same: the virulent mutant fails to become established. Van der Plank (1963) pointed out that the more widespread a new resistant variety becomes, the more selection for virulence will increase. As the crop approaches uniformity for resistance, only the virulent parasite strains will survive. If it were possible, between one season and the next, to replace completely all existing susceptible varieties with resistant ones, obligate pathogens like *Erysiphe* or *Puccinia* might be eliminated. No populations of spores could be formed in which virulent mutants would arise. However, such a scheme appears to be completely impractical for widely cultivated crops because it could never be enforced.

It seems very likely that in most parasite populations that are dense enough to cause economic loss, natural mutation rates are sufficient to generate mutants virulent on any nearby host protected by a single resistance gene. The suggestion that hosts should be protected by two or even three genes essentially buys time, since virulent mutants that arise independently must now be combined, or else they must arise simultaneously or sequentially in the same line before infection can occur.

Gene Frequencies—Selection

The fitness of a mutant allele in the absence of the resistant host also determines its frequency in the population. There is usually no selection for virulence on nonresistant hosts or, in the case of facultative parasites, some other substrate. If a virulent strain cannot compete with avirulent strains and other organisms under these circumstances, it may not survive to threaten the resistant variety. A number of authors have suggested that accumulation of virulence genes in a parasite will reduce fitness. Mutation to virulence involving deletion of parts of the parasite chromosome complement could be quite crippling. So could many point mutations. The result should be that selection stabilizes the frequency of virulence genes in the parasite population, eliminating those unnecessary for survival. Enough examples are known of virulence genes occurring frequently in parasite populations, where they seem not to be necessary, to show that by no means all virulence genes lower fitness (Van der Plank 1968; Watson 1970a; Luig and Watson 1970). Although a short-term experiment may show that one culture of a pathogen displaces another over the course of several consecutive generations of asexual spores, it is difficult to prove that this is due to the virulence genes being compared and not to other factors. A single isolate of a physiologic race may not be representative, especially if it has been in culture for some time. The fact remains that some virulence genes do seem to lower pathogen fitness. Van der Plank (1968) suggested that those resistance genes which, in a sense, impose a lowered fitness on parasite mutants that are virulent towards them be called "strong genes." I discussed their hypothetical characteristics on page 148. Resistance genes that have no such effect were called weak genes. Weak genes appear to be much more common than strong genes.

For obligate parasites we have an interesting paradox. The more widely a strong gene is used, the less it can be relied on for protection. The fitter races that can compete with races virulent on the strong gene are eliminated, and the crop selects the race that may ultimately destroy it. In the case of nonobligate parasites, stabilizing selection may well ensure that a virulent mutant with lowered fitness never survives on nonhost substrates to bring about reinfection. There are several examples of nonobligate plant diseases that have been effectively controlled by the introduction of a single gene for resistance, and where physiologic specialization has never been a problem. One is the gene for resistance to *Cercospora* blight in cucumber, carried originally by the English greenhouse variety "Butcher's Disease Resister." I am indebted to L. A. Darby and P. Grimbly for the following information. "Butcher's Disease Resister" was selected and bred by Bob Butcher of Dunstable, Bedfordshire, and released in 1903. It was derived from one of a few plants that survived the *Cercospora* epidemics prevalent on greenhouse crops during the period 1896–1907. Its introduction, together with soil sterilization that killed overwintering inoculum in the soil, helped eliminate the disease. Modern greenhouse cucumbers derive their resistance from this variety. In Europe *Cercospora* only attacks cucumbers in heated greenhouses, and there is no other source of inoculum than infected greenhouse crops. Modern practice also calls for lower temperatures and humidities, and as traditional greenhouses are replaced by modern, well ventilated houses, the possibility of the disease becoming serious again seems remote. Effective control of this disease depended on a combination of improved cultural methods and resistance. Specific resistance did not bear the entire burden of control and did not fail.

An extensive analysis of changes in race types in *Puccinia graminis tritici* was recently published by Luig and Watson (1970). Data from race surveys of this pathogen from three wheat regions of Australia, and from New Zealand, were examined to see how virulence had evolved over a period of approximately 50 years, a time span representing three eras. The first, extending from 1919, when records of rust variation began to be kept, to 1938, was a period when few if any resistant wheats were grown. The second, from 1938 to 1964, was marked by the common use of varieties with single genes for rust resistance. In the third era, from 1964 to the present, wheat varieties with three or more genes for resistance have been available.

Each period showed characteristic changes. During the first period, the rust races found in 1919, which were presumably adapted to indigenous grasses, were replaced by other races that possessed virulence genes not essential for their survival. For example, race 126–6,7, carrying virulence to *Sr8* and *Sr15,* first appeared in 1925 and remained dominant for 15 years. Luig and Watson suggested that this race originated outside Australia and was introduced, and that it was successful because of its greater fitness than native races. *Sr8* and *Sr15* are evidently weak genes.

During the second period there were frequent changes in race type, in response to changes in the resistant varieties grown. The alternate host of stem rust (*Berberis*) is so uncommon in Australia that sexual reproduction is virtually nonexistent. This was no handicap since variation occurred through mutation and mitotic recombination (see pages 70–72), and was probably supplemented, to a small extent, by exotic forms from overseas.

In the third period, the increased use of wheats with several genes for resistance was reflected by the increased recovery of correspondingly virulent rust races. The average number of virulence factors carried by isolates in the 5-year period 1964–68 was 3.18, whereas in the period 1954–58 it was only 1.46.

Although Luig and Watson's data provide some support for the concept of strong genes for rust resistance in wheat, these authors suggest that the lowered fitness of virulent mutants may be offset by response to selection. What may start out as a poorly adapted dikaryon homozygous for virulence and associated recessives may, as a result of further genetic change, become as fit as the rust type it replaces.

Comparable surveys of race changes in the North American populations of this rust fungus have been documented by Van der Plank (1968) and Watson (1970a) and similar trends recorded.

Specific Resistance—Robinson's Rules

The preceding pages may have created the impression that specific resistance is, with few exceptions, doomed to failure and consequently of little long term practical value. This of course is not the case. Specific resistance has long been, and continues to be, an effective way of protecting a great many crops. It may not often provide a final solution to a pest or disease problem, and for some diseases, potato blight being a case in point, it is of little or no value. Robinson (1971) examined the uses of this kind of disease resistance against a variety of plant pathogens with the object of establishing why it successfully controls some diseases and not others. His summary, drawn up as a set of rules, emphasizes that each disease has special features that distinguish it from many others, and that determine whether specific resistance will be effective. The following diseases will serve as examples. *Phytophthora infestans* is a parasite of potatoes. The host plants are polyploid and clonally propagated, and are consequently difficult to breed by the method of crossing an established variety with a source of resistance and of screening many thousands of seedlings for resistance and agronomic characters that match the susceptible parent. This means that rapid production of a succession of good agronomic varieties carrying different *R* genes comparable to rust resistant wheats is impractical. In favorable weather, epidemic spread of the pathogen is very rapid through airborne sporangia. Once established, a new race soon becomes widely distributed.

The wilt diseases of cabbage and tomato caused by *Fusarium oxysporum* are soil-borne and spread slowly, not by explosive outbursts caused by airborne spores. When crop rotation is practiced, physiologic races capable of infecting the resistant lines do not build up in the soil. Presumably the resistance genes are strong, in Van der Plank's sense, and selection eliminates the virulent races in the soil. Both cabbage and tomato are easy to breed, and good wilt-resistant varieties have achieved permanent control in spite of physiologic races. The same methods applied to control Panama disease of banana (*F. oxysporum cubense*) are not likely to succeed. The crop is difficult to breed and, as it is grown in plantations, rotation is impossible. New plantations are established by planting rhizome bits from older stock. These are likely to carry the wilt organism to the new plantation in the soil attached to the roots.

Multiline Varieties

Genetic uniformity in the host is an important factor in disease epidemics of crop plants. If heterogeneity can be deliberately introduced without compromising yield and quality, we could expect to gain some benefits. To take a simple example, if equal numbers of two resistant genotypes (*R1R1r2r2* and *r1r1R2R2*) were grown in a mixed stand in the presence of pathogen race 1 or race 2, but not both, we could anticipate at most a 50 percent loss. Experience has shown that release of the genotype *R1R1R2R2* selects pathogen race 1,2. Of course, race 1,2 could also destroy the mixed population, but the selection pressure for it is lower than on *R1R1R2R2*. Even if races 1 and 2 are present together, their build-up to epidemic proportions will be less rapid than that of race 1,2 on *R1R1R2R2*. This is because inoculum of race 1 that contacts *r1r1R2R2* is wasted, as is inoculum of race 2 on *R1R1r2r2*. Under some circumstances, for example when a grain crop is forming, a delay of only a few days could have a significant effect on crop yield.

Several workers have advocated the development of multiline varieties in which up to a dozen or more component lines are bulked. Each component line is produced by introducing resistance from a unique source into a common parent line by backcrossing. Selection ensures that the end products are similar in yield and maturity date, and that in the absence of disease, they are not inferior to a pure stand of the common parent. In theory, the composition of a mixed population can be varied in response to changes in the relative proportions of different virulent races of the pathogen. This depends on the availability of effective resistance genes from a number of different sources that have already been incorporated in the common varietal background and so are ready for use.

The multiline variety approach is well suited to breeding small grains. Its history, development, and theory were reviewed by Browning and Frey (1969). It has been carried furthest by oat breeders in Iowa and by the Rockefeller Foundation wheat breeding program in Colombia (Frey et al. 1971).

In practice, some variation in phenotype other than resistance is inevitable. The bulk of the oat crop is fed to livestock, and uniformity for such characters as milling, baking, or malting quality, which are used to judge other small grains, does not apply. Small grains for unsophisticated markets can also tolerate similar variation in grain quality. In fact, the background variation inherent in multiline varieties could be as important as the deliberately introduced variation in resistance. The reason for this is that the latter applies only to one, or possibly a few, parasites. It is therefore in a sense polarized. In all other respects, the multiline variety is as nearly uniform as the breeder can make it and is therefore just as vulnerable as the common parent to other parasites. However, uncontrolled variability could alter this situation and introduce useful, if unpredictable, resistance. I will return to this topic on page 192.

Regional Deployment of Resistance Genes

The seasonal progress of some parasites that cover large geographical areas involves the movement of spores or winged insects along set paths, determined by crop development and prevailing winds. These are parasites that do not survive locally between one season and the next but rather originate elsewhere. Three examples are the major oat pathogens *Puccinia graminis avenae, P. coronata avenae,* and the aphid-borne barley yellow-dwarf virus, all of which may be carried over great distances in North America by the wind. When specific resistance is deployed against such pathogens, there are likely to be advantages in not using the same resistance gene, or genes, throughout the whole geographic range of the crop. If different resistance genes are used where the disease overwinters from those where it does not, the continuity of susceptible hosts required for inoculum build-up and continued dispersal is missing. Browning et al. (1969) suggested two or three such zones for resistant oat varieties. However, the degree of cooperation necessary to ensure that 8 or 10 different genes for resistance are deployed in each of two zones, with no genes common to both zones, could be very difficult to achieve. In their view, a lower number of resistance genes would be inadequate and might enable a variable pathogen like crown rust to bridge the gap between zones. The chief difficulty is pressure on the local plant breeder to use the best genes for resistance that are available regardless of where else they may be in use.

Prescription or Profligacy

Extensive plant exploration and examination of exotic germ plasm has yielded many new and valuable genes for specific resistance to parasites. Understandably, in the past these genes have been pressed into service without much regard to the fact that they would probably be eventually rendered valueless by new virulent forms of the parasite they were to control. Released in combinations of two or three, or more, at a time, they seemed likely to survive for longer, since it became more difficult for the parasite to evolve virulence to all of them. Whether we are in danger of exhausting the supply of resistance genes is difficult to say. This question can only be answered by detailed examination of the germplasm resources of each crop. Even if there are plenty of resistance genes left, the cost of incorporating them into agronomic varieties is such that the question whether we can afford to squander them arises.

Antibiotics in medicine provide a partial analogy. Penicillin is an effective method of controlling some diseases caused by penicillin-sensitive bacteria. Mutants resistant to penicillin proved to be common, and by general agreement penicillin and other antibiotics were made available only by prescription, partly to discourage trivial uses that would accelerate the appearance of antibiotic resistant pathogens. Although there are no exactly comparable "trivial" uses of resistance to control plant parasites, it is now necessary to consider very carefully the value of short-term control measures, such as the release of varieties carrying a single gene for resistance, which by their failure would compromise more effective use of the same genes over a longer term, either in multilines, in multigenic varieties, or in a scheme of regional deployment.

Game Theory

The interaction between host and parasite in many respects resembles a game in which host and parasite make moves and countermoves. The host uses various strategies to minimize its losses, whereas the parasite uses strategies to maximize its gains. In natural populations the game rarely reaches a conclusion in which either host or parasite is eliminated. The dependence of the parasite on its host, which is at its most extreme for obligate parasites, regulates the effect of the parasite on the host. The outcome is usually a condition of equilibrium in which both components are assured of survival.

For most crop plants, man has interfered in an attempt to win the game outright and has usually failed. Game theory is likely to be helpful in visualizing the kinds of moves that will promote a compromise—a host-parasite equilibrium favorable to man. Such a plan may enable a plant breeder to manage the evolution of the parasite rather than be managed by it himself (Knott 1971).

General Resistance

The problems associated with specific resistance have prompted a good many plant breeders to explore control by general resistance. General resistance rarely provides complete control, but if a compromise is possible that provides a high enough level of stable resistance, a breeder may be prepared to include it despite the difficulties introduced by its multigenic inheritance. Perhaps the most important question is whether general resistance will last. Parasites are just as capable of polygenically determined variation as their hosts. Genetic studies of quantitative characters in fungi, such as growth rate or ascospore size in *Neurospora* (Papa 1971; Pateman and Lee 1960), and similar characters in insects, such as bristle number, fecundity, and so on, confirm the polygenic basis of their inheritance. Does this mean that a parasite, given sufficient time, could evolve genotypes capable of causing severe disease or damage to a crop protected by polygenic general resistance?

There are several ways of looking at this question. One is to suppose that each host polygene for resistance has an effect that is qualitatively, but not quantitatively, equivalent to that of a major gene for specific resistance. If the level of analysis is fine enough, the effect should be detectable. I would expect that it could be negated by changes in the parasite like those that occur for race-specific resistance. This interpretation defines general resistance as a form of specific resistance where the many individual effects are too small to be readily detected and specificity is consequently blurred. As appropriate virulence genes accumulate in the parasite, resistance will be gradually eroded.

Another view supposes that each polygene effect makes the host less attractive or less suitable as a substrate for the parasite, that there are very many different effects, and that they are essentially nonspecific in nature. These host genes could modify epidermal hair density, stomatal aperture, cuticle thickness, distribution and nature of mechanical tissues, availability of energy-rich metabolites, growth rate, day length responses, and so on. Their cumulative effect will be resistance to some parasites but not others. Resistance to different parasites will require other combinations. This type of resistance should last. From a teleological viewpoint, the several kinds of changes required of the parasite to negate it are either impractical or beyond its capacity for adaptation.

In practice, both specific and nonspecific polygenic effects may occur together in general resistance. The likely number of genes involved in rather high levels of general resistance to late blight in potatoes was discussed on page 31.

A good place to look for evidence of what has happened over a period of time is among crop varieties that have carried general resistance for many years. In potatoes, some selection for general resistance to late blight began following the late blight epidemics of the 1840's. Even so, the majority of existing commercial varieties in North America and Europe carry little or no resistance to late blight and must be protected by fungicides (Black 1970). In Holland, annual records of the tuber and foliage resistance ratings of potato varieties include a number with varying levels of general resistance. Van der Plank (1971) recently examined the records for 10 such varieties over a 30-year period from 1938 to 1968. He concluded that there was no evidence that resistance of any of the varieties had changed during the 30-year period in the way expected if they had selected blight races that were adapted to them. Although Van der Plank's conclusions are encouraging, they are, as he admits, based only on annual subjective ratings of the varieties concerned. They are accompanied by few statistics on the annual acreages of the varieties from which one might gauge the extent of genetic uniformity for this kind of blight resistance in the crop as a whole. Nor is there information on the extent to which blight was controlled by fungicides. Both are factors which would influence the evolution of blight races. The available laboratory evidence bearing on the question, although conflicting, does confirm that strains of *P. infestans* that grow rapidly and sporulate well on the tubers of one variety do not rapidly adapt to grow equally well on tubers of another (see Van der Plank 1971; Day 1966).

The tests of variety and inbred collections of wheat and corn to prevalent races of the rusts *P. striiformis* and *P. sorghi* discussed in Chapter 2 (pages 27 and 29) showed no evidence of erosion of general resistance. However, Johnson and Taylor (1972) recently described two forms of *P. striiformis* assigned to the same physiologic race. They showed small, but consistent, differences in their reactions on seedlings of three wheat varieties. For example on the variety Joss Cambier the reaction types were 3+ (much sporulation with barely detectable chlorosis) and 4 (much sporulation with no chlorosis) respectively. The second form produced twice the weight of uredospores on inoculated seedlings of Joss Cambier when compared with the first. Clearly such a change could account for the widely scattered and sometimes severe outbreaks of yellow rust reported for the first time on this winter wheat in England in 1971. While this is not an example of the erosion of general resistance it does demonstrate the kind of change in the pathogen that could lead to erosion.

Practical experience in controlling *P. sorghi* by selection of corn inbreds

with general resistance has so far shown no problems, and it has been practiced since hybrid corn began to be widely grown in North America in the
early 1940's. Other outstanding examples of general resistance include the
control of the bacterial wilt of tobacco and the peanut, both caused by
Pseudomonas solanacearum (Kelman 1953). Resistant tobacco was introduced in 1945 in North Carolina, and the peanut in the early 1920's in
Indonesia. Both releases followed intensive breeding and selection, and
both are still apparently as resistant today as they were originally.

Farmers and horticulturalists rely on general resistance to many
parasites to a much greater degree than they may be aware. The very
effective general resistance to *Ustilago maydis,* accumulated by rigorous
selection during the development of maize inbreds, is a case in point. This
resistance shows up when known highly susceptible lines are planted
alongside current commercial inbreds and hybrids. Under conditions of
natural infection, the former may have up to 75 percent or more smutted
plants, whereas galls on the latter are very infrequent. In countries where
U. maydis is now a problem, such as Poland and South Africa, the selection of resistant inbreds by introduction and accumulation of polygenes
in locally adapted material has not yet had time to be effective. In North
America, we tend to take smut resistance for granted.

Genetic Uniformity—A Technological Hazard

Plant breeding has played an important role in increasing agricultural
efficiency. Modern crop varieties deliver high yields of high quality food
and fiber. Many of the products must meet exacting industrial standards
for further processing: wheat for milling and baking; potatoes for frozen
french fries, chips, and soups; peas for freezing and canning; cotton of uniform fiber length, and so on. The emphasis on high yield and high return
has led inevitably to a great reduction in the genetic variation of all the
crops grown in intensive agricultures. In many cases the older, lower
yielding varieties, grown in former years, that together encompassed a
considerable part of the variation available among cultivated forms have
disappeared. In each major crop, they have been replaced by a small
number of more closely related "superior" forms.

No farmer can afford not to grow the highest yielding varieties that give
the greatest cash return. When there is only one best variety for a given
area, everyone will grow it and genetic uniformity will result. Uniformly
high yielding crops grown in dense stands in an intensive agriculture are
also uniformly vulnerable to any parasite that can exploit the opportunity
so provided. This is the very stuff of epidemics.

The trend has not gone unnoticed. Writing nearly 50 years ago, Jones

(1925) noted that extreme uniformity in the size of growth and time of flowering in corn hybrids could be disadvantageous if plants were all adversely affected by unfavorable weather at a critical time. With some variation, not all the plants would be damaged. It was some time later that examples of the same hazard in respect to parasite damage were pointed out by Hartley (1939) for clonally propagated crops, and by Stevens (1948) for these and self-pollinated crops. More recently Simmonds (1962) drew attention to the dangers of genetic uniformity and outlined some remedies. In 1967, the Food and Agricultural Organization (FAO) organized a conference in Rome to discuss genetic resources in crops and their conservation (Frankel and Bennet 1970). The most recent study followed in the wake of the Southern corn leaf blight epidemic in 1970.

The report of the National Academy committee on genetic vulnerability in major crops (Horsfall et al. 1972) showed how dangerously narrow the genetic base of many of the most important North American crops had become. In no case investigated were there grounds for complacency. The cause was clearly identified as the current emphasis on technological efficiency and maximal productivity.

Table 7.1, reproduced from the report, shows the quite small numbers of varieties that predominate in the acreages of the most important crops

TABLE 7.1
Acreage and farm value of major U.S. crops in 1969 and extent to which small numbers of cultivars dominate crop acreage. (From Horsfall et al. 1972.)

Crop	Acreage millions	Value millions	Total varieties	Major varieties number	Acreage %
Bean, dry	1.4	143	25	2	60
Bean, snap	0.3	99	70	3	76
Cotton	11.2	1,200	50	3	53
Corn[a]	66.3	5,200	197	6	71
Millet	2	?		3	100
Peanut	1.4	312	15	9	95
Peas	0.4	80	50	3	96
Potato	1.4	616	82	4	72
Rice	1.8	449	14	4	65
Soybean	42.4	2,500	62	6	56
Sugar beet	1.4	367	16	2	42
Sweet potato	0.13	63	48	1	69
Wheat	44.3	1,800	269	9	50

[a]Corn includes seeds, forage, and silage. Varieties refers only to use made of 197 released corn inbreds developed by Agricultural Experiment Stations.

grown in the United States. The same is true of major crops in Australia (Day 1973) and is probably a characteristic of advanced agricultures.

Other trends to genetic vulnerability were clearly evident. All hybrid sugar beet in the United States uses the same source of cytoplasmic male sterility. The same is true of sorghum. These are situations that will be remedied as soon as the search for alternative sterile cytoplasms is successful. Male-sterile barley and wheat in hybrid seed production plots are peculiarly susceptible to ergot (*Claviceps purpurea*) (Puranik and Mathre 1971). This disease infects flowers at the time of pollination. Ergot and other flower infections are not a problem if the period when flowers remain open is short. Flowers of sterile forms of cereals remain open a long time, greatly increasing the chances of their becoming infected. The continued popularity of the potato variety Russet Burbank (or Netted Gem) since 1873 was discussed on page 30. This variety is highly susceptible to potato blight and other pests and diseases and depends on pesticide applications for protection when epidemics threaten. It has some resistance to common scab (*Streptomyces scabies*) and Yellow Dwarf virus. Russet Burbank has outstanding qualities for baking and processing, and its tubers command high prices. It has proved to be very difficult to breed a variety with the same qualities but resistant to other pests and diseases. This is why potato farmers tolerate its vulnerability.

The impact of a major epidemic is greatly aggravated if agriculture is unprepared, and if the crop is the chief source of food for most of the population. Southern corn leaf blight in the United States could not have produced a famine like the potato blight in Ireland in the 1840's because alternative resistant corn hybrids were available and the society had other sources of carbohydrate and food for livestock.

Although the diversity of crops is an insurance against undue reliance on any one of them, the maintenance and development of alternative forms of individual crops that can be called on in case of need requires constant vigilance. Most epidemics are difficult if not impossible to anticipate. Who could have guessed the precise implications of introducing the gene *Pc-2* for crown rust resistance in oats before 1940, or *Tms* in corn before 1956, for two species of *Helminthosporium?* The assumption that exclusive use of one source of cytoplasmic male sterility in sugar beet is risky seems straightforward. But are there implications for such genes as "stringless" in beans, or "determinate" in tomatoes which are now uniformly distributed throughout these crops as a result of plant breeding? Could they carry a penalty like *Pc-2?* The many genes that are probably common to different genera and species have survived evolution and natural selection in a varying background of other genes before man's interference. This is not true of the genes on which agricultural technology now depends.

Future tomatoes which are uniform-ripening, high in vitamin C, free of cracks and greenback, and resistant to ozone and other atmospheric pollutants may be vulnerable to new forms of pests and diseases just as those with determinate plant habit were more liable to early blight caused by *Alternaria solani* (J. G. Horsfall: personal communication). A recently discovered, but minor, example of genetic vulnerability is the susceptibility of the apple variety Golden Delicious to a leaf blotch caused by the fungus *Glomerella cingulata.* Thompson and Taylor (1971) found that the susceptibility of Golden Delicious was due to a dominant gene *Gb* which was heterozygous. Threats of this kind cannot be precisely anticipated, but they can be prepared for by maintaining collections of germ plasm, and by developing alternative varieties that might be grown in place of vulnerable forms at short notice.

Multiline varieties represent the deliberate introduction of variability to counter a known threat from one parasite. They also have some potential for countering unpredicted threats. It is likely that not all interactions between the multiline variety and another parasite will have an identical outcome. This may be due either to pleiotropic effects of the different genes for resistance to the first parasite or to residual variability or to both. To this extent, multilines may better simulate natural variation than was originally anticipated. Current work on the productivity of mixed varietal stands of small grains may have much potential for other crops. As Allard and Hansche (1964) concluded, ". . . optimal yield may depend on breeding varieties in which the appropriate compromise is found between the demands for uniformity and the advantages of diversity."

BIBLIOGRAPHY

*The italicized figures in brackets following each reference indicate
the pages of this book where the work in question is mentioned.*

Adkisson, P. L. 1969. How insects damage crops, pp. 155–64. In *How crops grow* (P. R. Day, ed.). Conn. Agr. Exp. Sta. Bull. 708. [*156, 157*]

Aist, J. R., and P. H. Williams. 1972. Ultrastructure and time course of mitosis in the fungus *Fusarium oxysporum. J. Cell Biol.* 55:368–89. [*37, 38, 39*]

Allard, R. W. 1960. *Principles of plant breeding.* New York: Wiley. 485 pp. [*13, 32*]

————, and P. E. Hansche. 1964. Some parameters of population variability and their implications in plant breeding. *Advan. Agron.* 16:281–325. [*192*]

Allen, P. J. 1959. Physiology and biochemistry of defense, pp. 435–67. In *Plant pathology.* Vol. 1 (J. G. Horsfall and A. E. Dimond, eds.). New York: Academic. [*127*]

Anagnostakis, S. L. 1971. Cytoplasmic and nuclear control of an interstrain interaction in *Ustilago maydis. Mycologia* 63:94–97. [*80*]

Anderson, N. G. 1970. Evolutionary significance of virus infection. *Nature* 227:1346–47. [*153*]

Anglade, P., and P. M. Molot. 1967. Mise en évidence d'une relation entre les sensibilités de lignées de maïs à la pyrale (*Ostrinia nubilalis* Hbn) et à l'Helminthosporiose (*Helminthosporium turcicum* Pass). *Ann. Epiphyt.* 18:279–84. [*118*]

Anonymous. 1971. Bracts foil boll weevils. *Agr. Res.,* USDA 20 (6):3–4. [*162*]

Anonymous. 1972. Parasites and pesticides don't mix. *Agr. Res.,* USDA 20 (9):11. [*162*]

Antonelli, E., and J. M. Daly. 1966. Decarboxylation of indoleacetic acid by near-isogenic lines of wheat resistant or susceptible to *Puccinia graminis* f. sp. *tritici. Phytopathology* 56:610–18. [*127*]

Arnold, M. H., and S. J. Brown. 1968. Variation in the host-parasite relationship of a crop disease. *J. Agr. Sci.* 71:19–36. [*31*]

Ascher, K. R. S. 1970. Insect pest control by chemosterilants and antifeedants: Magdeburg 1966 to Milan 1969. *World Rev. Pest Control* 9:140–55. [*159*]

Audus, L. J. (ed.). 1964. *The physiology and biochemistry of herbicides.* London: Academic. 555 pp. [*171*]

Bagga, H. S., and D. M. Boone. 1968a. Genes in *Venturia inaequalis* controlling pathogenicity to crabapples. *Phytopathology* 58:1176–82. [*60, 106*]

———, and ———. 1968b. Inheritance of resistance to *Venturia* inaequalis in crabapples. *Phytopathology* 58:1183–87. [*105*]

Bailey, J. A., and B. J. Deverall. 1971. Formation and activity of phaseollin in the interaction between bean hypocotyls (*Phaseolus vulgaris*) and physiological races of *Colletotrichum lindemuthianum. Physiol. Plant Pathol.* 1:435–49. [*120*]

Baker, E. P. 1966. Isolation of complementary genes conditioning crown rust resistance in the oat variety Bond. *Euphytica* 15:313–18. [*22*]

———, and C. Teo. 1966. Mutants of *Puccinia graminis avenae* induced by ethyl methane sulphonate. *Nature* 209:632–33. [*68*]

Baker, K. F. and R. J. Cook. 1974. *Biological control of plant pathogens.* W. H. Freeman and Co., San Francisco. [*168*]

Baltimore, D. 1970. RNA-dependent DNA polymerase in virions of RNA tumour viruses. *Nature* 226:1209–11. [*153*]

Barrus, M. F. 1911. Variation in varieties of beans in their susceptibility to anthracnose. *Phytopathology* 1:190–95. [*81*]

Bartos, P., P. L. Dyck, and D. J. Samborski. 1969. Adult-plant leaf rust resistance in Thatcher and Marquis wheat: a genetic analysis of the host-parasite situation. *Can. J. Bot.* 47:267–69. [*12, 96*]

———, G. Fleischmann, D. J. Samborski, and W. A. Shipton. 1969. Studies on asexual variation in the virulence of oat rust, *Puccinia coronata* f. sp. *avenae* and wheat leaf rust, *Puccinia recondita. Can. J. Bot.* 47:1383–87. [*68, 70, 71, 72*]

Bartz, J. A., and J. E. Mitchell. 1970. Evidence for the metabolic detoxification of n-dodecylguanidine acetate by ungerminated macroconidia of *Fusarium solani* f. sp. *phaseoli. Phytopathology* 60:350–54. [*167*]

Bauch, R. 1930. Über multipolare Sexualität bei *Ustilago longissima. Arch. Protistenk.* 70:417–66. [*73, 74, 76*]

———. 1932. Die sexualität von *Ustilago scorzonereae* und *Ustilago zeae. Phytopathol. Z.* 5:315–21. [*74*]

Beadle, G. W., and E. L. Tatum. 1945. Neurospora II. Methods of producing and detecting mutations concerned with nutritional requirements. *Amer. J. Bot.* 32:678–86. [*37*]

Beck, S. D. 1965. Resistance of plants to insects. *Ann. Rev. Entomol.* 10:207–32. [*15*]

Bennett, C. W. 1969. Seed transmission of plant viruses. *Advan. Virus Res.* 14:221–61. [*34*]

Beraha, L., and E. D. Garber. 1971. Avirulence and extracellular enzymes of *Erwinia carotovora. Phytopathol. Z.* 70:335–44.   [*147*]

Beroza, M. 1972. Attractants and repellants for insect pest control, pp. 226–53. In *Pest control strategies for the future.* Washington, D.C.: Nat. Acad. Sci.   [*159*]

Biffen, R. H. 1905. Mendel's laws of inheritance and wheat breeding. *J. Agr. Sci.* 1:4–48.   [*10, 20*]

———. 1912. Studies in inheritance in disease resistance II. *J. Agr. Sci.* 4:421–29.   [*10*]

Bingefors, S. 1971. Resistance to nematodes and the possible value of induced mutations. *Mutation breeding for disease resistance.* IAEA (Vienna): 209–35.   [*15*]

Bingham, R. T., R. J. Hoff, and G. I. McDonald. 1971. Disease resistance in forest trees. *Ann. Rev. Phytopathol.* 9:433–52.   [*15*]

———, ———, ——— (eds.). 1972. *Biology of rust resistance in forest trees.* Proceedings of NATO-IUFRO Advanced Study Institute. USDA FS Misc. Publ. 1221. 681 pp.   [*15*]

Black, W. 1970. The nature and inheritance of field resistance to late blight (*Phytophthora infestans*) in potatoes. *Amer. Potato J.* 47:279–88.
                    [*15, 188*]

———, C. Mastenbroek, W. R. Mills, and L. C. Peterson. 1953. A proposal for an international nomenclature of races of *Phytophthora infestans* and of genes controlling immunity in *Solanum demissum* derivatives. *Euphytica* 2:173–79.   [*97*]

Boccas, B. 1972. Contribution a l'étude du cycle chez les *Phytophthora.* Analyse du mode de transmission d'un caractère génétique quantitatif chez une espèce homothallique, le *Phytophthora syringae* Kleb. *C. R. Acad. Sci.,* Sér. D (Paris), 275:663–66.   [*48*]

Bollen, G. J., and F. Scholten. 1971. Acquired resistance to benomyl and some other systemic fungicides in a strain of *Botrytis cinerea* in cyclamen. *Neth. J. Plant Pathol.* 77:83–90.   [*164*]

Bolley, L. H. 1905. Breeding for resistance or immunity to disease. *Proc. Amer. Breeders Ass.* 1:131–35.   [*15*]

Boone, D. M. 1971. Genetics of *Venturia inaequalis. Ann. Rev. Phytopathol.* 9:297–318.   [*61, 105, 124*]

———, and G. W. Keitt. 1956. *Venturia inaequalis* (Cke.) Wint. VIII. Inheritance of color mutant characters. *Amer. J. Bot.* 43:226–33.   [*62*]

———, and ———. 1957. *Venturia inaequalis* (Cke.) Wint. XII. Genes controlling pathogenicity of wild-type lines. *Phytopathology* 47:403–09.
                    [*60, 97, 106*]

Bourke, P. M. A. 1970. Use of weather information in the prediction of plant disease epiphytotics. *Ann. Rev. Phytopathol.* 8:345–70.   [*51*]

Bozarth, R. F. 1972. Mycoviruses: a new dimension in microbiology. *Environ. Health Perspect.* 2:23–39.   [*87*]

———, H. A. Wood, and R. R. Nelson. 1972. Viruslike particles in virulent strains of *Helminthosporium maydis. Phytopathology* 62:748 (abst.).   [*87*]

Bracker, C. E. 1968. Ultrastructure of the haustorial apparatus of *Erysiphe graminis* and its relationship to the epidermal cell of barley. *Phytopathology* 58:12–30. [55]

Brasier, C. M. 1971. Induction of sexual reproduction in single A^2 isolates of *Phytophthora* species by *Trichoderma viride. Nat. New Biol.* 231:283. [45]

Brefeld, O. 1883. Die Brandpilze I. (Ustilageen). *Bot. Untersuch. Hefenpilze* 5:1–219. [76]

Brinkerhoff, L. A. 1963. Variability of *Xanthomonas malvacearum:* the cotton bacterial blight pathogen. Okla. State Univ. Tech. Bull. T-98. 96 pp. [6]

———. 1970. Variation in *Xanthomonas malvacearum* and its relation to control. *Ann. Rev. Phytopathol.* 8:85–110. [97]

Britten, R. J., and E. H. Davidson. 1969. Gene regulation for higher cells: a theory. *Science* 165:349–57. [*132, 133*]

Brock, R. D. 1967. Disease resistance breeding. *J. Aust. Inst. Agr. Sci.* 33: 72–76. [*15*]

Brönnimann, A. 1968. Zur Toleranz des Weizens gegenüber *Septoria nodorum* Berk. *Phytopathol. Z.* 62:365–70. [*32*]

Browder, L. E. 1971a. Pathogenic specialization in cereal rust fungi, especially *Puccinia recondita* f. sp. *tritici:* concepts, methods of study, and application. USDA Tech. Bull. 1432. 51 pp. [*110*]

———. 1971b. A proposed system for coding infection types of the cereal rusts. *Plant Dis Rep.* 55:319–22. [*110*]

———. 1971c. The use of the computer in analysis of *Puccinia recondita* data. *Phytopathology* 61:886 (abst.). [*110*]

Brown, A. M. 1940. An aberrant strain of *Puccinia helianthi* Schw. *Can. J. Res. (C)* 18:513–17. [*68*]

———, and T. Johnson. 1949. Studies on variation in pathogenicity in leaf rust of wheat, *Puccinia triticina* Erikss. *Can. J. Res. (C)* 27:191–202. [*68*]

Browning, J. A., and K. J. Frey. 1969. Multiline cultivars as a means of disease control. *Ann. Rev. Phytopathol.* 7:355–82. [*15, 185*]

———, M. D. Simons, K. J. Frey, and H. C. Murphy. 1969. Regional deployment for conservation of oat crown-rust resistance genes, pp. 49–56. In *Disease consequences of intensive and extensive culture of field crops* (J. A. Browning, ed.). Iowa Agr. Exp. Sta. Spec. Rep. 64. [*185*]

Bryant, T. R., and K. L. Howard. 1969. Meiosis in the Oomycetes I. A microspectrophotometric analysis of nuclear DNA in *Saprolegnia terrestris* Cookson. *Amer. J. Bot.* 56:1075–83. [*46*]

Burges, H. D., and N. W. Hussey (eds.). 1971. *Microbial control of insects and mites.* New York: Academic. 861 pp. [*159, 163*]

Burk, L. G., and H. E. Heggestad. 1966. The genus *Nicotiana:* a source of resistance to diseases of cultivated tobacco. *Econ. Bot.* 20:76–88. [*15*]

Burrows, V. D. 1970. Absence of lateral transfer within the oat leaf of substances responsible for crown rust resistance. *Can. J. Bot.* 48:198–99. [*138*]

Bushnell, W. R., and D. M. Stewart. 1971. Development of American isolates of *Puccinia graminis* f. sp. *tritici* on an artificial medium. *Phytopathology* 61:376–79. [*64*]

Butcher, A. C. 1968. The relationship between sexual outcrossing and heterokaryon incompatibility in *Aspergillus nidulans*. *Heredity* 23:443–52.
[*41*]

Buxton, E. W. 1956. Heterokaryosis and parasexual recombination in pathogenic strains of *Fusarium oxysporum*. *J. Gen. Microbiol.* 15:133–39.
[*81*]

———. 1957. Differential rhizosphere effects of three pea cultivars on physiologic races of *Fusarium oxysporum* .f. *pisi*. *Trans. Brit. Mycol. Soc.* 40:305–17. [*168*]

Caldwell, R. W., W. B. Cartwright, and L. E. Compton. 1946. Inheritance of Hessian fly resistance derived from W38 and durum P.I. 94587. *J. Amer. Soc. Agron.* 38:398–409. [*102*]

Casida, J. E., and L. Lykken. 1969. Metabolism of organic pesticide chemicals in higher plants. *Ann. Rev. Entomol.* 20:607–36. [*171*]

Castro, J., G. A. Zentmyer, and W. L. Belser, Jr. 1971. Induction of auxotrophic mutants in *Phytophthora* by ultraviolet light. *Phytopathology* 61:283–89. [*46*]

Caten, C. E. 1971. Heterokaryon incompatibility in imperfect species of *Aspergillus*. *Heredity* 26:299–312. [*41*]

———. 1972. Vegetative incompatibility and cytoplasmic infection in fungi. *J. Gen. Microbiol.* 72:221–9. [*88*]

———, and J. L. Jinks. 1966. Heterokaryosis: Its significance in wild homothallic ascomycetes and fungi imperfecti. *Trans. Brit. Mycol. Soc.* 49:81–93. [*41*]

Chauhan, N. S. 1970. Genetic evidence of an unorthodox chromosomal system in the lac insect *Kerria lacca* (Kerr). *Genet. Res.* 16:341–44. [*102*]

Cheung, D. S. M., and H. N. Barber. 1972. Activation of resistance of wheat to stem rust. *Trans. Brit. Mycol. Soc.* 58:333–36. [*138*]

Chevaugeon, J. 1968. Étude expérimentale d'une étape du développement du *Pestalozzia annulata* B. et C. *Ann. Sci. Nat.* (Bot.) 9:417–32. [*87*]

———, and C. Lefort. 1960. Sur l'apparition régulière d'un "mutant" infectant chez un champignon du genre *Pestalozzia*. *C. R. Acad. Sci.* 250:2247–49. [*87*]

Clarke, D. D., and N. F. Robertson. 1966. Mutational studies on *Phytophthora infestans* (in relation to its adaptation to potato resistance). *Eur. Potato J.* 9:208–15. [*46*]

Cole, H., B. Taylor, and J. Duich. 1968. Evidence of differing tolerances to fungicides among isolates of *Sclerotinia homoeocarpa*. *Phytopathology* 58:683–86. [*164*]

Cotter, R. U., and B. J. Roberts. 1963. A synthetic hybrid of two varieties of *Puccinia graminis*. *Phytopathology* 53:344–46. [*69*]

Cousin, R. 1965. Étude de la résistance a l'oïdium chez les pois. *Ann. Amélior. Plantes* 95:93–97. [*20*]

Couture, R. M., D. G. Routley, and G. M. Dunn. 1971. Role of cyclic hydroxamic acids in monogenic resistance of maize to *Helminthosporium turcicum*. *Physiol. Plant Pathol.* 1:515–21. [*119*]

Craigie, J. H. 1959. Nuclear behavior in the diploidization of haploid infections of *Puccinia helianthi. Can. J. Bot.* 37:843–54. [*68*]

Cruickshank, I. and D. Perrin. 1960. Isolation of a phytoalexin from *Pisum sativum* L. *Nature* 187:799–800. [*119*]

_____, and _____. 1963. Phytoalexins of the Leguminosae. Phaseollin from *Phaseolus vulgaris* L. *Life Sci.* 2:680–82. [*119*]

DaCosta, C. P., and C. M. Jones. 1971. Cucumber beetle resistance and mite susceptibility controlled by the bitter gene in *Cucumis sativus* L. *Science* 172:1145–46. [*123*]

Daft, M. J., and W. D. P. Stewart. 1971. Bacterial pathogens of freshwater blue-green algae. *New Phytol.* 70:819–29. [*172*]

Daly, J. M. 1972. The use of near-isogenic lines in biochemical studies of the resistance of wheat to stem rust. *Phytopathology* 62:392–400. [*117*]

_____, P. Ludden, and P. M. Seevers. 1971. Biochemical comparisons of resistance to wheat stem rust disease controlled by the *Sr6* or *Sr11* alleles. *Physiol. Plant Pathol.* 1:397–407. [*128, 131*]

Damian, R. T. 1964. Molecular mimicry: antigen-sharing by parasite and host and its consequences. *Amer. Nat.* 98:129–50. [*151*]

Darwin, C. 1868. *The variation of animals and plants under domestication.* Vol. 1. London: Murray. 354 pp. [*9*]

Davies, J. M. L., and D. G. Jones. 1970. The origin of a diploid "hybrid" of *Cercosporella herpotrichoides. Heredity* 25:137–39. [*46*]

Davis, R. H. 1966. Heterokaryosis, pp. 567–88. In *The fungi: an advanced treatise.* Vol. 2 (G. C. Ainsworth and A. S. Sussman, eds.). New York: Academic. [*41*]

Day, A. W. 1972. Genetic implications of current models of somatic nuclear division in fungi. *Can. J. Bot.* 50:1337–47. [*39*]

_____, and J. K. Jones. 1968. The production and characteristics of diploids in *Ustilago violacea. Genet. Res.* 11:63–81. [*74*]

_____, and _____. 1969. Sexual and parasexual analysis of *Ustilago violacea. Genet. Res.* 14:195–221. [*78*]

Day, B. E., et al. 1968. Weed control. In *Principles of plant and animal pest control.* Vol. 2. Washington, D.C.: Nat. Acad. Sci. 471 pp. [*173*]

Day, P. R. 1956. Race names of *Cladosporium fulvum. Tomato Genet. Co-op. Rep.* 6:13–14. [*97, 107*]

_____. 1957. Mutation to virulence in *Cladosporium fulvum. Nature* 179:1141–42. [*141, 142*]

_____. 1960. Variation in phytopathogenic fungi. *Ann. Rev. Microbiol.* 14:1–16. [*3, 97*]

_____. 1966. Recent developments in the genetics of the host-parasite system. *Ann. Rev. Phytopathol.* 4:245–68. [*188*]

_____. 1968. Plant disease resistance. *Sci. Progr.* 56:357–70. [*15, 169*]

_____. 1972a. The genetics of rust fungi, pp. 3–17. In *Biology of rust resistance in forest trees* (R. T. Bingham, R. J. Hoff, and G. I. McDonald, eds.). USDA FS Misc. Publ. 1221. [*67, 68*]

_____. 1972b. Crop resistance to pests and pathogens, pp. 257–71. In *Pest control strategies for the future.* Washington, D.C.: Nat. Acad. Sci. [*15*]

_____. 1973. Genetic variability of crops. *Ann. Rev. Phytopathol.* 11:293–312. [179, 191]

_____, and S. L. Anagnostakis. 1971a. Corn smut dikaryon in culture. *Nat. New Biol.* 231:19–20. [76]

_____, and _____. 1971b. Meiotic products from natural infections of *Ustilago maydis. Phytopathology* 61:1020–21. [75, 78]

_____, _____. 1973. The killer system in *Ustilago maydis.* Heterokaryon transfer and loss of determinants. *Phytopathology* 63:1017–18. [80, 87]

_____, _____, and J. E. Puhalla. 1971. Pathogenicity resulting from mutation at the *b* locus of *Ustilago maydis. Proc. Nat. Acad. Sci.* 68:533–35. [75, 141]

_____, and K. J. Scott. 1973. Scanning electron microscopy of fresh material of *Erysiphe graminis* f. sp. *hordei. Physiol. Plant Pathol.* 3:433–35. [56]

Dekker, J. 1969. Acquired resistance to fungicides. *World Rev. Pest Contr* 8:79–85. [166, 167]

_____. 1971. Selective action of fungicides and development of resistance in fungi to fungicides. *Proc. 6th. Brit. Insectic. Fungic. Conf.:* 715–23. [164]

Denward, T. 1970. Differentiation in *Phytophthora infestans* II. Somatic recombination in vegetative mycelium. *Hereditas* 66:35–48. [50, 51]

DeVay, J. E., R. Charudattan, and D. L. S. Wimalajeewa. 1972. Common antigenic determinants as a possible regulator of host-pathogen compatibility. *Amer. Nat.* 106:185–94. [149]

Dewit-Elshove, A., and A. Fuchs. 1971. The influence of the carbohydrate source on pisatin breakdown by fungi pathogenic to pea (*Pisum sativum*). *Physiol. Plant Pathol* 1:17–24. [122]

Dineen, J. K. 1963. Antigenic relationship between host and parasite. *Nature* 197:471–72. [151]

Dinoor, A., J. Khair, and G. Fleischmann. 1968a. Pathogenic variability and the unit representing a single fertilization in *Puccinia coronata* var. *avenae. Can. J. Bot.* 46:501–08. [66, 68]

_____, _____, _____. 1968b. A single-spore analysis of rust pustules produced by mixing isolates of two races of *Puccinia coronata var. avenae. Can. J. Bot.* 46:1455–58. [66]

Doane, C. C. 1971. Field application of a *Streptococcus* causing brachyosis in larvae of *Porthetria dispar. J. Invert. Pathol.* 17:303–07. [159]

D'Oliveira, B. 1940. Notes on the production of the aecidial stage of some cereal rusts in Portugal. *Rev. Agron.* (Lisbon) 28:201–08. [66, 68]

Dorn, G. L. 1967. A revised map of the eight linkage groups of *Aspergillus nidulans. Genetics* 56:619–31. [62]

Doubly, J. A., H. H. Flor, and C. O. Clagget. 1960. Relation of antigens of *Melampsora lini* and *Linum usitatissimum* to resistance and suceptibility. *Science* 131:229. [149]

Doy, C. H., P. M. Greshoff, and B. G. Rolfe. 1973. Biological and molecular evidence for transgenosis of genes from bacteria to plant cells. *Proc. Nat. Acad. Sci.* 70:723–26. [152]

Dunn, G. M., and T. Namm. 1970. Gene dosage effects on monogenic resistance to northern corn leaf blight. *Crop Sci.* 10:352–54. [20]

Duran, R., and S. M. Norman. 1961. Differential sensitivity to biphenyl among strains of *Penicillium digitatum*. *Plant Dis. Rep.* 45:475–80.   [*164*]

Durham, W. F., and C. H. Williams. 1972. Mutagenic, teratogenic, and carcinogenic properties of pesticides. *Ann. Rev. Entomol.* 17:123–48.
[*155*]

Duvick, D. N. 1965. Cytoplasmic pollen sterility in corn. *Advan. Genet.* 13:1–56.   [*34, 35*]

Dyck, P. L., and D. J. Samborski. 1968. Genetics of resistance to leaf rust in the common wheat varieties Webster, Loris, Brevit, Carina, Malakof and Centenario. *Can. J. Genet. Cytol.* 10:7–17.   [*21*]

Dyer, T. A., and K. J. Scott. 1972. Decrease in chloroplast content of barley leaves infected with powdery mildew. *Nature.* 236:237–38.   [*131*]

Dyte, C. E. 1967. Possible new approach to the chemical control of plant feeding insects. *Nature* 216:298.   [*159, 169*]

Ehrlich, M. A., and H. G. Ehrlich. 1971. Fine structure of the host-parasite interfaces in mycoparasitism. *Ann. Rev. Phytopathol.* 9:155–84.   [*124*]

Ellingboe, A. H. 1965. Somatic recombination in Basidiomycetes, pp. 36–48. In *Incompatibility in fungi* (K. Esser and J. R. Raper, eds.). Berlin: Springer.   [*70*]

———. 1972. Genetics and physiology of primary infection by *Erysiphe graminis*. *Phytopathology* 62:401–06.   [*129, 130*]

Elliott, C. G., and D. MacIntyre. 1973. Genetical evidence on the life-history of *Phytophthora*. *Trans. Brit. Mycol. Soc.* 60:311–16.   [*49*]

Epstein, S. S., and M. S. Legator. 1971. *The mutagenicity of pesticides: concepts and evaluation.* Cambridge, Mass.: MIT. 220 pp.   [*155, 170*]

Eshed, N., and I. Wahl. 1970. Host ranges and interrelations of *Erysiphe graminis hordei, E. graminis tritici* and *E. graminis avenae*. *Phytopathology* 60:628–34.   [*57*]

Esposito, R. E. (née R. Easton), and R. Holliday. 1964. The effect of 5-fluorodeoxyuridine on genetic replication and somatic recombination in synchronously dividing cultures of *Ustilago maydis*. *Genetics* 50:1009–17.
[*78*]

Estrada, R., N. and J. Guzman N. 1969. Herencia de la resistencia de campo al "tizon" (*Phytophthora infestans* Mont. de Bary). En variedades cultivadas de papa (subspecies *tuberosa* y *andigena*) *Rev. Inst. Colombiano Agropec.* 4:117–37.   [*31*]

Falcon, L. A. 1971. Use of bacteria for microbial control, pp. 67–95. In *Microbial control of insects and mites* (H. D. Burges and N. W. Hussey, eds.). New York: Academic.   [*159*]

Falconer, D. S. 1960. *Introduction to quantitative genetics.* New York: Ronald. 365 pp.   [*32*]

Favret, E. A. 1965. Induced mutations in breeding for disease resistance. *The use of induced mutation in plant breeding.* (Suppl. to *Radiat. Bot.* 5):521–36.
[*16*]

———. 1971. The host-pathogen system and its genetic relationships, pp. 457–71. In *Barley genetics II: Proc. 2nd Int. Barley Genet. Symp.* (R. A. Nilan, ed.) Pullman, Wn.: Washington State Univ. Press.   [*16, 21, 147*]

Fenner, F. 1965. Myxoma virus and *Oryctolagus cuniculus:* two colonizing species, pp. 485–99. In *Genetics of colonizing species* (H. G. Baker and G. L. Stebbins, eds.). New York: Academic. [*163*]

Férault, A. C., D. Spire, F. Rapilly, J. Bertrandy, M. Skajennikoff, and P. Bernaux. 1971. Observation de particules virales dans des souches de *Piricularia oryzae* Briosi et Cav. *Ann. Phytopathol.* 3:267–69. [*87*]

Fincham, J. R. S., and P. R. Day. 1971. *Fungal genetics.* 3rd ed. Oxford: Blackwell; Philadelphia: Davis. 402 pp. [*24, 37, 55, 62, 144, 180*]

Fischer, G. W., and C. S. Holton. 1957. *Biology and control of the smut fungi.* New York: Ronald. 622 pp. [*73*]

Flangas, A. L., and J. G. Dickson. 1961. The genetic control of pathogenicity, serotypes and variability in *Puccinia sorghi. Amer. J. Bot.* 48:275–85. [*96*]

Fleischmann, G., and R. I. H. McKenzie. 1968. Inheritance of crown rust resistance in *Avena sterilis. Crop Sci.* 8:710–13. [*14*]

Flor, H. H. 1942. Inheritance of pathogenicity in *Melampsora lini. Phytopathology* 32:653–69. [*66, 94, 96*]

――――. 1946. Genetics of pathogenicity in *Melampsora lini. J. Agr. Res.* 73:335–57. [*67, 94*]

――――. 1947. Inheritance of reaction to rust in flax. *J. Agr. Res.* 74:241–62. [*22, 94*]

――――. 1955. Host-parasite interaction in flax rust: its genetics and other implications. *Phytopathology* 45:680–85. [*95, 142*]

――――. 1956a. The complementary genic systems in flax and flax rust. *Adv. Genet.* 8:29–54. [*68, 95*]

――――. 1956b. Mutations in flax rust induced by ultraviolet radiation. *Science* 124:888–89. [*141, 144*]

――――. 1957. The vegetative origin of a new race of the flax rust fungus. *Phytopathology* 47:11 (abst.). [*68*]

――――. 1958. Mutation to wider virulence in *Melampsora lini. Phytopathology* 48:297–301. [*141, 144*]

――――. 1959. Differential host range of the monocaryon and dicaryons of a eu-autoecious rust. *Phytopathology* 49:794–95. [*67*]

――――. 1960a. Asexual variants of *Melampsora lini. Phytopathology* 50:223–26. [*68*]

――――. 1960b. The inheritance of X-ray-induced mutations to virulence in a urediospore culture of race 1 of *Melampsora lini. Phytopathology* 50:603–05. [*68, 142, 143*]

――――. 1964. Genetics of somatic variation for pathogenicity in *Melampsora lini. Phytopathology* 54:823–26. [*68, 70*]

――――. 1965a. Tests for allelism of rust resistance genes in flax. *Crop Sci.* 5:415–18. [*22, 26*]

――――. 1965b. Inheritance of smooth-spore-wall and pathogenicity in *Melampsora lini. Phytopathology* 55:724–27. [*68*]

――――. 1971. Current status of the gene-for-gene concept. *Ann. Rev. Phytopathol.* 9:275–96. [*143, 144, 147*]

――――, and V. E. Comstock. 1971. Development of flax cultivars with multiple rust-conditioning genes. *Crop Sci.* 11:64–66. [*22*]

Frank, J. A., and J. D. Paxton. 1970. Time sequence for phytoalexin production in Harosoy and Harosoy 63 soybeans. *Phytopathology* 60:315–18.
[*120*]

———, and ———. 1971. An inducer of soybean phytoalexin and its role in the resistance of soybeans to *Phytophthora* rot. *Phytopathology* 61: 954–58.
[*120*]

Frankel, O. H., and E. Bennett. 1970. *Genetic resources in plants: their exploration and conservation.* IBP Handbook 11. Philadelphia: Davis. 554 pp.
[*190*]

Frey, K. J., J. A. Browning, and R. L. Grindeland. 1971. Implementation of oat multiline cultivar breeding. *Mutation breeding for disease resistance.* IAEA (Vienna):159–69.
[*185*]

Frost, L. C. 1961. Heterogeneity in recombination frequencies in *Neurospora crassa. Genet. Res.* 2:43–62.
[*62*]

Galindo, J., and M. E. Gallegly. 1960. The nature of sexuality in *Phytophthora infestans. Phytopathology* 50:123–28.
[*45*]

———, and G. A. Zentmyer. 1967. Genetical and cytological studies of *Phytophthora* strains pathogenic to pepper plants. *Phytopathology* 57: 1300–04.
[*46*]

Gallegly, M. E. 1970. Genetics of *Phytophthora. Phytopathology* 60:1135–41.
[*48*]

———, and J. J. Eichenmuller. 1959. The spontaneous appearance of the potato race 4 character in cultures of *Phytophthora infestans. Amer. Potato J.* 36:45–51.
[*51*]

Gallun, R. L., H. O. Deay, and W. B. Cartwright. 1961. Four races of Hessian fly selected and developed from an Indiana population. Purdue Univ. Agr. Exp. Sta. Res. Bull. 732:1–8.
[*100, 101*]

———, and J. H. Hatchett. 1969. Genetic evidence of elimination of chromosomes in the Hessian fly. *Ann. Entomol. Soc. Amer.* 62:1095–1101.
[*102*]

———, and L. P. Reitz. 1971. Wheat cultivars resistant to races of Hessian fly. USDA Prod. Res. Rep. 134. 16 pp.
[*100, 104*]

Garber, E. D. 1960. The host as a growth medium. *Ann. N. Y. Acad. Sci.* 88: 1187–94.
[*146*]

Georghiou, G. P. 1972. The evolution of resistance to pesticides. *Ann. Rev. Ecol. Syst.* 3:133–68.
[*157, 158*]

Georgopoulos, S. G. 1963. Tolerance to chlorinated nitrobenzenes in *Hypomyces solani* f. *cucurbitae* and its mode of inheritance. *Phytopathology* 53:1086–93.
[*166*]

———. 1964. Chlorinated-nitrobenzene tolerance in *Sclerotium rolfsii. Ann. Inst. Phytopathol. (Benaki),* n.s. 6:156–59.
[*164*]

———, and H. D. Sisler. 1970. Gene mutation eliminating antimycin A-tolerant electron transport in *Ustilago maydis. J. Bact.* 103:745–50.
[*166*]

———, and C. Zaracovitis. 1967. Tolerance of fungi to organic fungicides. *Ann. Rev. Phytopathol.* 5:109–30.
[*165, 166*]

Gerhold, H. D., R. E. McDermott, E. J. Schreiner, and J. A. Winieski (eds.). 1966. *Breeding pest-resistant trees.* Oxford: Pergamon. 505 pp.
[*15*]

Giatgong, P., and R. A. Frederiksen. 1969. Pathogenic variability and cytology of monoconidial subcultures of *Piricularia oryzae*. *Phytopathology* 59:1152–57.   [*85, 86*]

Gibson, R. W. 1971. Glandular hairs providing resistance to aphids in certain wild potato species. *Ann. Appl. Biol.* 68:113–19.   [*112*]

Gilbert, L. E. 1971. Butterfly-plant coevolution: Has *Passiflora adenopoda* won the selectional race with Heliconiine butterflies? *Science* 172:585–86.
                                                                            [*112*]

Girbardt, M. 1965. Eine Zielschnittmethode für Pilzzellen. *Mikroskopie* 20:254–64.   [*37*]

Goldschmidt, V. 1928. Verebungsversuche mit den biologischen Arten des Antherenbrandes (*Ustilago violacea*) Ein Beitrag zur Frage der parasitären spezialisierung. *Z. Bot.* 21:1–90.   [*43, 110*]

Goodman, R. N. 1972. Phytotoxin-induced ultrastructural modifications of plant cells, pp. 311–28. In *Phytotoxins in plant diseases* (R. K. S. Wood, A. Ballio, and A. Graniti, eds.), London: Academic.   [*124*]

Graham, K. M. 1963. Inheritance of resistance to *Phytophthora infestans* in two diploid Mexican *Solanum* species. *Euphytica* 12:35–40.   [*31*]

Green, G. J. 1963. Stem rust of oats in Canada in 1963. *Can. Plant Dis. Surv.* 43:173–78.   [*108*]

———. 1964. A color mutation, its inheritance, and the inheritance of pathogenicity in *Puccinia graminis* Pers. *Can. J. Bot.* 42:1653–64.   [*67, 96*]

———. 1965. Stem rust of wheat, rye and barley in Canada in 1964. *Can. Plant Dis. Surv.* 45:23–29.   [*108*]

———. 1966. Selfing studies with races 10 and 11 of wheat stem rust. *Can. J. Bot.* 44:1255–60.   [*96*]

———. 1971a. Physiologic races of wheat stem rust in Canada from 1919 to 1969. *Can. J. Bot.* 49:1575–88.   [*108*]

———. 1971b. Hybridization between *Puccinia graminis tritici* and *Puccinia graminis secalis* and its evolutionary implications. *Can. J. Bot.* 49:2089–95.   [*69*]

———, and T. Johnson. 1958. Further evidence of resistance in *Berberis vulgaris* to race 15B of *Puccinia graminis tritici*. *Can. J. Bot.* 36:351–55.
                                                                             [*66*]

———, and M. A. S. Kirmani. 1969. Somatic segregation in *Puccinia graminis* f. sp. *avenae*. *Phytopathology* 59:1106–08.   [*68*]

———, and R. I. H. McKenzie. 1967. Mendelian and extrachromosomal inheritance of virulence in *Puccinia graminis* f. sp. *avenae*. *Can. J. Genet. Cytol.* 9:785–93.   [*68, 69, 87*]

Green, T. R., and C. A. Ryan. 1972. Wound-induced proteinase inhibitor in plant leaves: a possible defense mechanism against insects. *Science* 175:776–77.   [*123*]

Greenaway, W. 1971. Relationship between mercury resistance and pigment production in *Pyrenophora avenae*. *Trans. Brit Mycol. Soc.* 56:37–44.
                                                                            [*164*]

Grente, J. 1971. Hypovirulence et lutte biologique dans le cas de l'*Endothia parasitica*. *Ann. Phytopathol.* 3:409–10 (abst.).   [*87*]

————, and S. Sauret. 1969. L'hypovirulence exclusive, phénomène original en pathologie végétale. *C. R. Acad. Sci.*, Sér. D (Paris), 268:2347–50.
[*87, 88*]

Griffiths, D. J., and A. J. H. Carr. 1961. Induced mutation for pathogenicity in *Puccinia coronata avenae. Trans. Brit. Mycol. Soc.* 44:601–07.   [*68, 141*]

Grümmer, G., E. Günther, and D. Eggert. 1969. Die Prüfung von Tomatensorten und ihren Hybriden auf Blatt- und Fruchtbefall mit *Phytophthora infestans. Theor. Appl. Genet.* 39:232–38.   [*31*]

Habgood, R. M. 1970. Designation of physiological races of plant pathogens. *Nature* 227:1268–69.   [*109*]

Hadwiger, L. A., and M. E. Schwochau. 1969. Host resistance responses: an induction hypothesis. *Phytopathology* 59:223–27.   [*120*]

————, and ————. 1971. Specificity of deoxyribonucleic acid intercalating compounds in the control of phenylalanine ammonia lyase and pisatin levels. *Plant Physiol.* 47:346–51.   [*121*]

Halisky, P. M. 1965. Physiologic specialization and genetics of the smut fungi III. *Bot. Rev.* 31:114–50.   [*73, 77*]

Hamilton, R. H. 1964. A corn mutant deficient in 2,4-dihydroxy-7-methoxy-1,4-benzoxazin-3-one with an altered tolerance of atrazine. *Weeds* 12:27–30.   [*119*]

Hankin, L., and J. E. Puhalla. 1971. Nature of a factor causing interstrain lethality in *Ustilago maydis. Phytopathology* 61:50–53.   [*79*]

Harding, P. R. 1959. Biphenyl-induced variations in Citrus blue mold. *Plant Dis. Rep.* 43:649–53.   [*164*]

Harland, S. C. 1948. Inheritance of immunity to mildew in Peruvian forms of *Pisum sativum. Heredity* 2:263–69.   [*20*]

Hartl, D. L. 1969. Dysfunctional sperm production in *Drosophila melanogaster* males homozygous for the segregation distorter elements. *Proc. Nat. Acad. Sci.* 63:782–89.   [*161*]

Hartley, C. 1939. The clonal variety of tree planting: asset or liability? *Phytopathology* 29:9 (abst.).   [*190*]

Hartley, M. J., and P. G. Williams. 1971. Genotypic variation within a phenotype as a possible basis for somatic hybridization in rust fungi. *Can. J. Bot.* 49:1085–87.   [*70*]

Hastie, A. C. 1970a. The genetics of asexual phytopathogenic fungi with special reference to *Verticillium,* pp. 55–62. In *Root diseases and soil-borne pathogens* (T. A. Tousson, R. V. Bega, and P. E. Nelson, eds.). Berkeley: Univ. Calif.   [*81, 83*]

————. 1970b. Benlate-induced instability of *Aspergillus* diploids. *Nature* 226:771.   [*170*]

————, and S. G. Georgopoulos. 1971. Mutational resistance to fungitoxic benzimidazole derivatives in *Aspergillus nidulans. J. Gen. Microbiol.* 67:371–73.   [*166*]

Hatchett, J. H. 1969. Race E, sixth race of the Hessian fly, *Mayetiola destructor,* discovered in Georgia wheat fields. Ann. Entomol. Soc. Amer. 62:677–8.   [*102*]

———, and R. L. Gallun. 1970. Genetics of the ability of the Hessian fly, *Mayetiola destructor*, to survive on wheats having different genes for resistance. *Ann. Entomol. Soc. Amer.* 63:1400–07.   [*97, 100, 102, 103, 104*]

Heath, M. C., and I. B. Heath. 1971. Ultrastructure of an immune and a susceptible reaction of cowpea leaves to rust infection. *Physiol. Plant Pathol.* 1:277–87.   [*124*]

Heimpel, A. M. 1967. A critical review of *Bacillus thuringiensis* var. *thuringiensis* Berliner and other crystalliferous bacteria. *Ann. Rev. Entomol.* 12:287–322.   [*159*]

Heringa, R. J., A. Van Norel, and M. F. Tazelaar. 1969. Resistance to powdery mildew (*Erysiphe polygoni D. C.*) in peas (*Pisum sativum L.*) *Euphytica* 18:163–69.   [*20*]

Hermansen, J. E. 1959. Split pycnial lesions of *Puccinia graminis*. A study of spreading pycniospores including spores from pycnia of different color. *Friesia* 6:33–36.   [*63*]

Heumann, W. 1968. Conjugation in starforming *Rhizobium lupini*. *Mol. Gen. Genet.* 102:132–44.   [*152*]

Higgins, V. J., and R. L. Millar. 1969a. Comparitive abilities of *Stemphylium botryosum* and *Helminthosporium turcicum* to induce and degrade a phytoalexin from alfalfa. *Phytopathology* 59:1493–99.   [*121*]

———, and ———. 1969b. Degradation of alfalfa phytoalexin by *Stemphylium botryosum*. *Phytopathology* 59:1500–06.   [*121*]

Hilu, H. M., and A. L. Hooker, 1963. Monogenic chlorotic lesion resistance to *Helminthosporium turcicum* in corn seedlings. *Phytopathology* 53:909–12.   [*119*]

Hiura, U. 1965. Compatibility between form species of *Erysiphe graminis* DC and the pathogenicity of the spores formed after hybridization between form species. *Nogaku Kenkyu* 51:67–73 (see *Plant Breed. Abst.* 37:5960).   [*57*]

Hoffman, J. A., and E. L. Kendrick. 1969. Genetic control of compatibility in *Tilletia controversa*. *Phytopathology* 59:79–83.   [*73*]

Holcomb, R. W. 1970. Insect control: alternatives to the use of conventional pesticides. *Science* 168:456–58.   [*159, 163*]

Holliday, R. 1961. Induced mitotic crossing-over in *Ustilago maydis*. *Genet. Res.* 2:231–48.   [*75*]

———. 1964. The induction of mitotic recombination by mitomycin C in *Ustilago* and *Saccharomyces*. *Genetics* 50:323–35.   [*78*]

Hollings, M., and O. M. Stone. 1971. Viruses that infect fungi. *Ann. Rev. Phytopathol.* 9:93–118.   [*87*]

Holton, C. S., and P. M. Halisky. 1960. Dominance of avirulence and monogenic control of virulence in race hybrids of *Ustilago avenae*. *Phytopathology* 50:766–70.   [*96*]

———, J. A. Hoffman, and R. Duran. 1968. Variation in the smut fungi. *Ann. Rev. Phytopathol.* 6:213–42.   [*73, 77, 96*]

Hooker, A. L. 1967. The genetics and expression of resistance in plants to rusts of the genus *Puccinia*. *Ann. Rev. Phytopathol.* 5:163–82.   [*15*]

_____. 1969. Widely based resistance to rust in corn. In *Disease consequences of intensive and extensive culture of field crops* (J. A. Browning, ed.). Iowa Agr. Exp. Sta. Spec. Rep. 64:28–34. [15]

_____, and W. A. Russell. 1962. Inheritance of resistance to *Puccinia sorghi* in six corn inbred lines. *Phytopathology* 52:122–28. [96]

_____, and K. M. S. Saxena. 1971. Genetics of disease resistance in plants. *Ann. Rev. Genet.* 5:407–24. [15, 20]

_____, D. R. Smith, S. M. Lim, and M. D. Musson. 1970. Physiological races of *Helminthosporium maydis* and disease resistance. *Plant Dis. Rep.* 54:1109–10. [34, 177]

Horsfall, J. G., et al. 1972. *Genetic vulnerability of major crops.* Report to Nat. Acad. Sci., U.S.A. 307 pp. [15, 30, 35, 179, 190]

Hotson, H. H., and V. M. Cutter, Jr. 1951. The isolation and culture of *Gymnosporangium juniperi-virginianae* Schw. *Proc. Nat. Acad. Sci.* 37:400–03. [64]

Hougas, R. W., and S. J. Peloquin. 1958. The potential of potato haploids in breeding and genetic research. *Amer. Potato J.* 35:701–07. [30]

Hovanitz, W. 1969. Inherited and/or conditioned changes in host-plant preference in *Pieris. Entomol. Exp. Appl.* 12:729–35. [123]

Howard, H. W. 1968. The relation between resistance genes in potatoes and pathotypes of potato-root eelworm (*Heterodera rostochiensis*), wart disease (*Synchytrium endobioticum*) and potato virus X. *Absts. 1st Int. Congr. Plant Pathol.* (London):92. [97]

Hsiao, T. H. 1969. Chemical basis of host selection and plant resistance in oligophagous insects. *Entomol. Exp. Appl.* 12:777–88. [123]

Ingham, J. L. 1972. Phytoalexins and other natural products as factors in plant disease resistance. *Bot. Rev.* 38:343–424. [115]

Ingram, R. 1968. *Verticillium dahliae* var. *longisporum,* a stable diploid. *Trans. Brit. Mycol. Soc.* 51:339–41. [41]

Iwasaki, S., S. Nozoe, S. Okuda, Z. Sato, and T. Kozaka. 1969. Isolation and structural elucidation of a phytotoxic substance produced by *Pyricularia oryzae* Cavara. *Tetrahedron Lett.* 45:3977–80. [82]

Jinks, J. L. 1966. Extranuclear inheritance, pp. 619–60. In *The fungi: an advanced treatise,* Vol. 2 (G. C. Ainsworth and A. S. Sussman, eds.). New York: Academic. [42]

Johnson, R., R. W. Stubbs, E. Fuchs, and N. H. Chamberlain. 1972. Nomenclature for physiologic races of *Puccinia striiformis* infecting wheat. *Trans. Brit. Mycol. Soc.* 58:475–80. [109]

Johnson, R. and A. J. Taylor. 1972. Isolates of *Puccinia striiformis* collected in England from the wheat varieties Maris Beacon and Joss Cambier. *Nature,* 238, 105–6. [188]

Johnson, T. 1946. Variation and the inheritance of certain characters in rust fungi. *Cold Spring Harbor Symp. Quant. Biol.* 11:85–93. [69, 86, 87]

_____. 1949. Inheritance of pathogenicity and urediospore color in crosses between physiologic races of oat stem rust. *Can. J. Res. (C)* 27:203–17. [68, 69]

_____. 1953. Variation in the rusts of cereals. *Biol. Rev.* 28:105–57.   [*43*]

_____. 1954. Selfing studies with physiological races of wheat stem rust, *Puccinia graminis* var. *tritici. Can. J. Bot.* 32:506–22.   [*69, 87*]

_____. 1961. Man-guided evolution in plant rusts. *Science* 133:357–62.   [*110*]

_____, G. J. Green, and D. J. Samborski. 1967. The world situation of the cereal rusts. *Ann. Rev. Phytopathol.* 5:183–200.   [*108*]

Jones, D. F. 1925. *Genetics in plant and animal improvement.* New York: Wiley. 568 pp.   [*189*]

Jones, F. G. W., J. M. Carpenter, D. M. Parrott, A. R. Stone, and D. L. Trudgill. 1970. Potato cyst nematode: one species or two? *Nature* 227:83–84.   [*90*]

_____, and D. M. Parrott. 1965. The genetic relationship of pathotypes of *Heterodera rostochiensis* Woll. which reproduce on hybrid potatoes with genes for resistance. *Ann. Appl. Biol.* 56:27–36.   [*90, 97*]

Jones, G. A., and R. Thurston. 1970. Effect of an area program using black-light traps to control populations of tobacco hornworms and tomato hornworms in Kentucky. *J. Econ. Entomol.* 63:1186–94.   [*159*]

Jørgensen, J. H. 1971. Comparison of induced mutant genes with spontaneous genes in barley conditioning resistance to powdery mildew. In *Mutation breeding for disease resistance.* IAEA (Vienna):117–24.   [*16*]

Käfer, E. 1961. The processes of spontaneous recombination in vegetative nuclei of *Aspergillus nidulans. Genetics* 46:1581–1609.   [*81*]

Kao, K. N., and D. R. Knott. 1969. The inheritance of pathogenicity in races 111 and 29 of wheat stem rust. *Can. J. Genet. Cytol.* 11:266–74.   [*68, 96*]

Kappas, A., and S. G. Georgopoulos. 1970. Genetic analysis of dodine resistance in *Nectria haematococca* (syn. *Hypomyces solani*). *Genetics* 66:617–22.   [*166*]

_____, and _____. 1971. Independent inheritance of avirulence and dodine resistance in *Nectria haematococca* var. *cucurbitae. Phytopathology* 61:1093–94.   [*166*]

Keen, N. T., and R. Horsch. 1972. Hydroxyphaseollin production by various soybean tissues: a warning against use of "unnatural" host-parasite systems. *Phytopathology* 62:439–42.   [*120*]

_____, J. J. Sims, D. C. Erwin, E. Rice, and J. E. Partridge. 1971. 6α-hydroxyphaseollin: an antifungal chemical induced in soybean hypocotyls by *Phytophthora megasperma* var. *sojae. Phytopathology* 61:1084–89.   [*120*]

Kehr, A. E. 1966. Current status and opportunities for the control of nematodes by plant breeding. In *Pest control by chemical, biological, genetic and physical means* (E. F. Knipling, ed.). USDA 33–110:126–138.   [*15*]

Keiding, J. 1967. Persistence of resistant populations after the relaxation of the selection pressure. *World Rev. Pest Contr.* 6:115–30.   [*158*]

Keitt, G. W. 1952. Inheritance of pathogenicity in *Venturia inaequalis* (Cke.) Wint. *Amer. Nat.* 86:373–90.   [*60*]

————, and M. H. Langford. 1941. *Venturia inaequalis* (Cke.) Wint. I. A groundwork for genetic studies. *Amer. J. Bot.* 28:805–20.   [*59, 60*]

————, C. Leben, and J. R. Shay. 1948. *Venturia inaequalis* (Cke.) Wint. IV. Further studies on the inheritance of pathogenicity. *Amer. J. Bot.* 35:334–36.   [*61*]

Kelman, A. 1953. The bacterial wilt caused by *Pseudomonas solanacearum.* N. C. AES. Tech. Bull. 99. 194 pp.   [*183*]

————, and L. Sequeira. 1972. Resistance in plants to bacteria. *Proc. Roy. Soc.,* B (London), 181:247–66.   [*15*]

Kerr, A. 1971. Acquisition of virulence by non-pathogenic isolates of *Agrobacterium radiobacter. Physiol. Plant Pathol.* 1:241–46.   [*152*]

Kimber, G., and M. S. Wolfe. 1966. Chromosome number of *Erysiphe graminis. Nature* 212:318–19.   [*55*]

Király, Z., B. Barna, and T. Érsek. 1972. Hypersensitivity as a consequence, not the cause, of plant resistance to infection. *Nature* 239:456–58.   [*169*]

Klement, Z., and R. N. Goodman. 1967. The hypersensitive reaction to infection by bacterial plant pathogens. *Ann. Rev. Phytopathol.* 5:17–44.
[*15*]

Knight, T. A. 1799. *Phil. Trans. Roy. Soc.* :192.   [*9*]

Knipling, E. F. 1955. Possibilities of insect control or eradication through the use of sexually sterile males. *J. Econ. Entomol.* 48:459–62.   [*160*]

Knott, D. R. 1971. Can losses from wheat stem rust be eliminated in North America. *Crop Sci.* 11:97–99.   [*187*]

————, and R. G. Anderson. 1956. The inheritance of rust resistance. I. The inheritance of stem rust resistance in ten varieties of common wheat. *Can. J. Agr. Sci.* 36:174–95.   [*21*]

Koltin, Y., R. Kenneth, and I. Wahl. 1963. Powdery mildew disease of barley in Israel and physiologic specialization of the pathogen. *Proc. 1st Int. Barley Genet. Symp.* (Wageningen, Neth.) 1963:228–35.   [*93*]

Konzak, C. F. 1956. Induction of mutations for disease resistance in cereals. *Brookhaven Symp. Biol.* 9:141–56.   [*18*]

Kozar, F. 1969. Mitotic recombination in biochemical mutants of *Ustilago hordei. Can. J. Genet. Cytol.* 11:961–66.   [*78*]

Krieger, R. I., P. P. Feeny, and C. F. Wilkinson. 1971. Detoxication enzymes in the guts of caterpillars: an evolutionary answer to plant defenses? *Science* 172:579–81.   [*122*]

Kuć, J. 1966. Resistance of plants to infectious agents. *Ann. Rev. Microbiol.* 20:337–70.   [*15*]

————. 1972. Phytoalexins. *Ann. Rev. Phytopathol.* 10:207–32.
[*115, 117, 118, 124*]

Kuhl, J. L., D. J. Maclean, K. J. Scott, and P. G. Williams. 1971. The axenic culture of *Puccinia* species from uredospores: experiments on nutrition and variation. *Can. J. Bot.* 49:201–09.   [*64*]

Kuiper, J. 1965. Failure of hexachlorobenzene to control common bunt of wheat. *Nature* 206:1219–20.   [*164*]

Kuo, M., O. C. Yoder, and R. P. Scheffer. 1970. Comparative specificity of the toxins of *Helminthosporium carbonum* and *Helminthosporium victoriae*. *Phytopathology* 60:365–68. [53]

Langford, A. N. 1937. The parasitism of *Cladosporium fulvum* Cooke and the genetics of resistance to it. *Can. J. Res.* (C) 15:108–28. [5]

———. 1948. Autogenous necrosis in tomatoes immune from *Cladosporium fulvum* Cooke. *Can. J. Res.* (C) 26:35–64. [139]

Lapierre, H., J. M. Lemaire, B. Jouan, and G. Molin. 1970. Mise en évidence de particules virales associées à une perte de pathogénicité chez le Piétin— échaudage des céréales, *Ophiobolus graminis* Sacc. *C. R. Acad. Sci.*, Sér. D (Paris), 271:1833–36. [87, 89]

Latterell, F. M. 1972. Two views of pathogenic stability in *Pyricularia oryzae*. *Phytopathology* 62:771. Abstr. [85]

Lawes, D. A., and J. D. Hayes. 1965. The effect of mildew (*Erysiphe graminis* f. sp. *avenae*) on spring oats. *Plant Pathol.* 14:125–28. [55]

Ledoux, L., R. Huart, and M. Jacobs. 1971. Fate of exogenous DNA in *Arabidopsis thaliana* translocation and integration. *Eur. J. Biochem.* 23:96–108. [151]

Lemaire, J. M., B. Jouan, B. Perraton, and M. Sailly. 1971. Perspectives du lutte biologique contre les parasites des céréales d'origine tellurique en particulier *Ophiobolus graminis* Sacc. *Sci. Agron. Rennes* 1971:1–8.
[87, 89]

Leonard, K. J. 1971. Association of virulence and mating type among *Helminthosporium maydis* isolates collected in 1970. *Plant Dis. Rep.* 55:759–60. [178]

———. 1973. Association of mating type and virulence in *Helminthosporium maydis,* and observations on the origin of the Race T population in the United States. *Phytopathology.* 63:112–15. [178]

Leppik, E. E. 1970. Gene centers of plants as sources of disease resistance. *Ann. Rev. Phytopathol.* 8:323–44. [13]

Lewellen, R. T., E. L. Sharp, and E. R. Hehn. 1967. Major and minor genes in wheat for resistance to *Puccinia striiformis* and their responses to temperature changes. *Can. J. Bot.* 45:2155–72. [96]

Lewis, C. M., and G. M. Tarrant. 1971. Induction of mutation by 5-fluoroura-cil and amino acid analogues in *Ustilago maydis*. *Mut. Res.* 12:349–56.
[42]

Lewis, D. 1949. Structure of the incompatibility gene II. Induced mutation rate. *Heredity* 3:339–55. [140]

———. 1952. Serological reactions of pollen incompatibility substances. *Proc. Roy. Soc.,* 140:127–35. [150]

———, S. Burrage, and D. Walls. 1967. Immunological reactions of single pollen grains, electrophoresis and enzymology of pollen protein exudates. *J. Exp. Bot.* 18:371–78. [150]

Lhoas, P. 1961. Mitotic haploidization by treatment of *Aspergillus niger* diploids with parafluorophenylalanine. *Nature* 190:744. [42]

———. 1967. Genetic analysis by means of the parasexual cycle in *Aspergillus niger*. *Genet. Res.* 10:45–61. [81]

————. 1971a. Transmission of double stranded RNA viruses to a strain of *Penicillium stoloniferum* through heterokaryosis. *Nature* 230:248–49.

[*80*]

————. 1971b. Infection of protoplasts from *Penicillium stoloniferum* with double-stranded RNA viruses. *J. Gen. Virol.* 13:365–67.   [*80, 87*]

Lim, S. M., and A. L. Hooker. 1971. Southern corn leaf blight: genetic control of pathogenicity and toxin production in race T and race O of *Cochliobolus heterostrophus*. *Genetics* 69:115–17.   [*54*]

Lindberg, G. D. 1959. A transmissible disease of *Helminthosporium victoriae*. *Phytopathology* 49:29–32.   [*87*]

————. 1968. A symptomless carrier of disease in *Helminthosporium victoriae*. *J. Gen. Microbiol.* 50:361–65.   [*87*]

————. 1969. Separation of a mildly diseased from a severely diseased isolate of *Helminthosporium victoriae*. *Phytopathology* 59:1884–88.   [*87*]

————. 1971. Disease-induced toxin production in *Helminthosporium oryzae*. *Phytopathology* 61:420–24.   [*55, 87, 89*]

————, and T. P. Pirone. 1963. Serological differentiation of normal and diseased *Helminthosporium victoriae*. *Phytopathology* 53:881 (abst.).   [*87*]

Line, R. F., E. L. Sharp, and R. L. Powelson. 1970. A system for differentiating races of *Puccinia striiformis* in the United States. *Plant Dis. Rep.* 54:992—94.   [*96, 108*]

Linskens, H. F. 1960. Zur Frage der Enstehung der Abwehr-Körper bei der Inkompatibilitäts reaktion von *Petunia* III. *Z. Bot.* 48:126–35.   [*150*]

Little, R., and J. G. Manners. 1969a. Somatic recombination in yellow rust of wheat (*Puccinia striiformis*) I. The production and possible origin of two new physiologic races. *Trans. Brit. Mycol. Soc.* 53:251–58.   [*68, 70*]

————, and ————. 1969b. Somatic recombination in yellow rust of wheat (*Puccinia striiformis*) II. Germ tube fusions, nuclear number and nuclear size. *Trans. Brit. Mycol. Soc.* 53:259–67.   [*68, 70*]

Littlefield, L. J. 1969. Flax rust resistance induced by prior inoculation with an avirulent race of *Melampsora lini*. *Phytopathology* 59:1323–28.   [*135*]

————, and S. J. Aronson. 1969. Histological studies of *Melampsora lini* resistance in flax. *Can. J. Bot.* 47:1713–17.   [*124*]

————, and C. J. Bracker. 1972. Ultrastructural specialization at the host-pathogen interface in rust-infected flax. *Protoplasma* 74:271–305.   [*126*]

Locke, S. B. 1969. Botran tolerance of *Sclerotium cepivorum* isolants from fields with different Botran-treatment histories. *Phytopathology* 59:13 (abst.).   [*164*]

Loegering, W. Q. 1966. The relationship between host and pathogen in stem rust of wheat. Proc. 2nd Int. Wheat Genetics Symp. (Lund, 1963). *Hereditas,* Suppl. 2:167–77.   [*4*]

————, and L. E. Browder. 1971. A system of nomenclature for physiologic races of *Puccinia recondita tritici*. *Plant Dis. Rep.* 55:718–22.   [*108*]

————, and J. R. Geiss. 1957. Independence in the action of three genes conditioning stem rust resistance in Red Egyptian wheat. *Phytopathology* 47:740–41.   [*127*]

_____, and D. L. Harmon. 1969. Wheat lines near-isogenic for reaction to *Puccinia graminis tritici. Phytopathology* 59:456–59. [*127*]

_____, _____, and W. A. Clark. 1966. Storage of urediospores of *Puccinia graminis tritici* in liquid nitrogen. *Plant Dis. Rep.* 50:502–06. [*6, 63*]

_____, R. A. McIntosh, and C. H. Burton. 1971. Computer analysis of disease data to derive hypothetical genotypes for reaction of host varieties to pathogens. *Can. J. Genet. Cytol.* 13:742–48. [*105*]

_____, and H. R. Powers, Jr. 1962. Inheritance of pathogenicity in a cross of physiological races 111 and 36 of *Puccinia graminis* f. sp. *tritici. Phytopathology* 52:547–54. [*96*]

Loprieno, N. 1964. I mutanti nutrizionali nello studio dei rapporti ospite-patogeno nelle fitopatie da microorganismi. *Agr. Ital.,* Marzo–Aprile:1–15. [*146*]

Lovrekovich, L., H. Lovrekovich, and M. A. Stahmann. 1967. Inhibition of phenol oxidation by *Erwinia carotovora* in potato tuber tissue and its significance in disease resistance. *Phytopathology* 57:737–42. [*122*]

Lüers, H. 1953. Untersuchungen zur Frage der Mutagenizität des Kontaktinsektizids DDT an *Drosophila melanogaster. Naturwissenschaften* 40:293. [*170*]

Luig, N. H. 1967. A high reversion rate for yellow urediospore color in *Puccinia graminis* f. sp. *tritici. Phytopathology* 57:1091–93. [*68, 141*]

_____, and E. P. Baker. 1956. A note on a uredospore colour mutant in barley leaf rust. *Puccinia hordei* Otth. *Proc. Linn. Soc. N.S.W.* 81:115–18. [*68*]

_____, and I. A. Watson. 1961. A study of inheritance of pathogenicity in *Puccinia graminis* var. *tritici. Proc. Linn. Soc. N.S.W.* 86:217–29. [*96*]

_____, and _____. 1970. The effect of complex genetic resistance in wheat on the variability of *Puccinia graminis* f. sp. *tritici. Proc. Linn. Soc. N.S.W.* 95:22–45. [*181, 182*]

Luke, H. H., and V. E. Gracen, Jr. 1972. *Helminthosporium* toxins, Ch. 6, pp 139–68. In *Microbial toxins,* Vol. 8 *Fungal Toxins* (S. Kadis, A. Ciegler, and S. J. Ajl, eds). New York: Academic. [*53, 113*]

_____, H. C. Murphy, and F. C. Petr. 1966. Inheritance of spontaneous mutations of the Victoria locus in oats. *Phytopathology* 56:210–12. [*18*]

_____, and A. T. Wallace. 1969. Sensitivity of induced mutants of an *Avena* cultivar to victorin at different temperatures. *Phytopathology* 59:1769–70. [*18, 19*]

_____, H. E. Wheeler, and A. T. Wallace. 1960. Victoria-type resistance to crown rust separated from susceptibility to *Helminthosporium* blight in oats. *Phytopathology* 50:205–09. [*17*]

Lukezic, F. L., and J. E. DeVay. 1964. Effect of myo-inositol in host tissues on the parasitism of *Prunus domestica* var. President by *Rhodosticta quercina. Phytopathology* 54:697–700. [*146*]

Lupton, F. G. H., and R. Johnson. 1970. Breeding for mature-plant resistance to yellow rust in wheat. *Ann. Appl. Biol.* 66:137–43. [*26, 27, 29*]

_____, and R. C. F. Macer. 1962. Inheritance of resistance to yellow rust

(*Puccinia glumarum* Erikss. and Henn.) in seven varieties of wheat. *Trans. Brit. Mycol. Soc.* 45:21–45.   [*20*]

Macer, R. C. F. 1967. The occurrence of a virulent and genetically stable physiologic race of *Purrinia striiformis*. *Trans. Brit. Mycol. Soc.* 50:305–10.   [*68*]

Maclean, D. J., and K. J. Scott. 1970. Variant forms of saprophytic mycelium grown from uredospores of *Puccinia graminis* f. sp. *tritici*. *J. Gen. Microbiol.* 64:19–27.   [*64, 65*]

MacNeill, B. H., and G. L. Barron. 1966. Avirulence in prototrophs of *Penicillium expansum*. *Can. J. Bot.* 44:355–58.   [*146*]

Madelin, M. F. 1966. Fungal parasites of insects. *Ann. Rev. Entomol.* 11:423–48.   [*159*]

Magaich, B. B., M. D. Upadhya, O. Prakosh, and S. J. Singh. 1968. Cytoplasmically determined expression of symptoms of potato virus x in crosses between species of *Capsicum*. *Nature* 220:1341–42.   [*34*]

Malcolmson, J. F. 1969. Races of *Phytophthora infestans* occurring in Great Britain. *Trans. Brit. Mycol. Soc.* 53:417–23.   [*49*]

————. 1970. Vegetative hybridity in *Phytophthora infestans*. *Nature* 225:971–72.   [*50*]

Malone, J. P. 1968. Mercury-resistant *Pyrenophora avenae* in Northern Ireland seed oats. *Plant Pathol.* 17:41–45.   [*164*]

Manibhushanrao, K. M. & P. R. Day. 1972. Low night temperature and blast disease development on rice. *Phytopathology* 62:1005–7.   [*85*]

Mann, B. 1962. Role of pectic enzymes in the *Fusarium* wilt syndrome of tomato. *Trans. Brit. Mycol. Soc.* 45:169–78.   [*146*]

Mansfield, J. W., and B. J. Deverall. 1971. Mode of action of pollen in breaking resistance of *Vicia faba* to *Botrytis cinerea*. *Nature* 232:339.   [*122*]

Martens, J. W., R. I. H. McKenzie, and G. J. Green. 1970. Gene-for-gene relationships in the *Avena: Puccinia graminis* host-parasite system in Canada. *Can. J. Bot.* 48:969–75.   [*96*]

Matta, A. 1971. Microbial penetration and immunization of uncongenial host plants. *Ann. Rev. Phytopathol.* 9:387–410.   [*116*]

Maxwell, F. G., J. N. Jenkins, and W. J. Parrott. 1972. Resistance of plants to insects. *Adv. Agron.* 24:187–265.   [*15*]

McClanahan, R. J. 1970. Integrated control of the greenhouse whitefly on cucumbers. *J. Econ. Entomol.* 63:599–601.   [*159*]

McIntosh, R. A., and E. P. Baker. 1969. Chromosome location and linkage studies involving the *Pm3* locus for powdery mildew resistance in wheat. *Proc. Linn. Soc. N.S.W.* 93:232–38.   [*136, 138*]

McKee, R. K. 1969. Effects of ultraviolet irradiation on zoospores of *Phytophthora infestans*. *Trans. Brit. Mycol. Soc.* 52:281–91.   [*46, 141*]

McLaughlin, R. E. 1971. Use of protozoans for microbial control of insects, Ch. 6, pp. 151–72. In *Microbial control of insects and mites* (H. D. Burges and N. W. Hussey, eds.). New York: Academic.   [*159*]

Mercado, A. C., Jr., and R. M. Lantican. 1961. The susceptibility of cytoplasmic male sterile lines of corn to *Helminthosporium maydis* Nish. and Miy. *Phil. Agr.* 45:235–43.   [*34, 177*]

Metzger, R. J., and E. J. Trione. 1962. Application of the gene-for-gene relationship hypothesis to the *Triticum-Tilletia* system. *Phytopathology* 52:363 (abst.).   [96]

Miah, M. A. J. 1968. Method of selfing F1 cultures and interpreting genetic data in rust fungi. *Can. J. Genet. Cytol.* 10:613–19.   [66]

———, and W. E. Sackston. 1970. Genetics of pathogenicity in sunflower rust. *Phytoprotection* 51:17–35.   [68, 96]

Michewicz, J. E., D. L. Sutton, and R. D. Blackburn. 1972. The White Amur for aquatic weed control. *Weed Sci.* 20:106–10.   [172]

Miles, P. W. 1969. Interaction of plant phenols and salivary phenolases in the relationship between plants and Hemiptera. *Entomol. Exp. Appl.* 12:736–44.   [122]

Moreland, D. E. 1967. Mechanisms of action of herbicides. *Ann. Rev. Plant Physiol.* 18:365–86.   [171]

Moseman, J. G. 1957. Host-parasite interactions between culture 12A1 of the powdery mildew fungus and the *Mlk* and *Mlg* genes in barley. *Phytopathology* 47:453 (abst.).   [96]

———. 1959. Host-pathogen interaction of the genes for resistance in *Hordeum vulgare* and for pathogenicity in *Erysiphe graminis* f. sp. *hordei*. *Phytopathology* 49:469–72.   [96]

———. 1966. Genetics of powdery mildews. *Ann. Rev. Phytopathol.* 4:269–90.   [15, 57]

———, and H. R. Powers, Jr. 1957. Function and longevity of cleistothecia of *Erysiphe graminis* f. sp. *hordei*. *Phytopathology* 47:53–56.   [57]

Muldrew, J. A. 1953. The natural immunity of the larch sawfly (*Pristiphora erichsonii* (Htg)) to the introduced parasite *Mesoleius tenthredinis* Morley, in Manitoba and Saskatchewan. *Can. J. Zool.* 31:313–32.   [163]

Muller, C. H. 1970. The role of allelopathy in the evolution of vegetation, pp. 13–31. In *Biochemical co-evolution* (K. L. Chambers, ed.). Corvallis: Oregon State Univ. Press.   [171]

Müller, K. O. 1953. The nature of resistance of the potato plant to blight— *Phytophthora infestans*. *J. Nat. Inst. Agr. Bot.* 6:346–60.   [115]

———. 1956. Einige einfache versuchenzum nachweis von Phytoalexinen. *Phytopathol. Z.* 27:237–54.   [119]

———, and H. Börger. 1940. Experimentelle Untersuchungen über die Phytophthora: Resistenz der Kartoffel. *Arb. biol. Abt.* (Aust. Reichsanst.) (Berlin) 23:189–231.   [115]

Murray, M. J. 1969. Successful use of irradiation breeding to obtain *Verticillium* resistant strains of peppermint *Mentha piperita* L. *Induced mutations in plants.* IAEA (Vienna): 345–71.   [16]

Nasrallah, M. E., J. T. Barber, and D. H. Wallace. 1969. Self-incompatibility proteins in plants: detection, genetics, and possible mode of action. *Heredity* 25:23–27.   [150]

Neal, J. L., Jr., T. G. Atkinson, and R. I. Larson. 1970. Changes in the rhizosphere microflora of spring wheat induced by disomic substitution of a chromosome. *Can. J. Microbiol.* 16:153–58.   [168]

Nelson, R. R., J. E. Ayers, H. Cole, and D. H. Petersen. 1970. Studies and observations on the past occurrence and geographical distribution of isolates of race T of *Helminthosporium maydis*. *Plant Dis. Rep.* 54:1123–26.   [*6, 178*]

———, and D. M. Kline. 1968. Occurrence in *Cochliobolus heterostrophus* of capacities to blight gramineous hosts. *Plant Dis. Rep.* 52:879–82.   [*178*]

———, and ———. 1969. Genes for pathogenicity in *Cochliobolus heterostrophus*. *Can. J. Bot.* 47:1311–14.   [*53*]

Newton, M., and T. Johnson. 1927. Color mutations in *Puccinia graminis tritici* (Pers.) Erikss. and Henn. *Phytopathology* 17:711–25.   [*67*]

———, ———, and A. M. Brown. 1930. A study of the inheritance of spore colour and pathogenicity in crosses between physiologic forms of *Puccinia graminis tritici*. *Sci. Agr.* 10:775–98.   [*43, 67*]

Niederhauser, J. S. 1956. The blight, the blighter and the blighted. *Trans. N. Y. Acad. Sci.*, Ser. II 19:55–63.   [*93*]

———. 1968. Resistance to *Phytophthora infestans* in Mexico. *Abst. 1st Int. Congr. Plant Pathol:* 138.   [*51*]

Nielsen, J. 1968. Isolation and culture of monokaryotic haplonts of *Ustilago nuda*, the role of proline in their metabolism, and the inoculation of barley with resynthesized dikaryons. *Can. J. Bot.* 46:1193–1200.   [*73*]

Noble, W. B., and C. A. Suneson. 1943. Differentiation of the two genetic factors for resistance to the Hessian fly in Dawson wheat. *J. Agr. Res.* 67:27–32.   [*104*]

Noronha-Wagner, M., and A. J. Bettencourt. 1967. Genetic study of the resistance of *Coffea* spp. to leaf rust. 1. Identification and behaviour of four factors conditioning disease reaction in *Coffea arabica* to twelve physiologic races of *Hemileia vastatrix*. *Can. J. Bot.* 45:2021–31.   [*96*]

Nutman, P. S. 1969. Genetics of symbiosis and nitrogen fixation in legumes. *Proc. Roy. Soc.* 172:417–37.   [*97*]

Oort, A. J. P. 1963. A gene-for-gene relationship in the *Triticum-Ustilago* system and some remarks on host-parasite combinations in general. *Neth. J. Plant Pathol.* 69:104–09.   [*96*]

Oppenoorth, F. J. 1965. Biochemical genetics of insecticide resistance. *Ann. Rev. Entomol.* 10:185–206.   [*158*]

Orton, W. A. 1900. The wilt disease of cotton and its control. USDA Div. Veg. Physiol. Pathol. Bull. 27.   [*14*]

———. 1905. Plant breeding as a factor in controlling plant diseases. *Proc. Amer. Breeders Ass.* 1:69–72.   [*10*]

Ou, S. H., and M. R. Ayad. 1968. Pathogenic races of *Pyricularia oryzae* originating from single lesions and monoconidial cultures. *Phytopathology* 58:179–82.   [*82, 85*]

———, and P. R. Jennings. 1969. Progress in development of disease-resistant rice. *Ann. Rev. Phytopathol.* 7:383–410.   [*15, 85*]

Owens, L. D. 1969. Toxins in plant disease: structure and mode of action. *Science* 165:18–24.   [*53*]

Painter, R. H. 1951. *Insect resistance in crop plants.* Macmillan: New York. 520 pp. (Annotated reprint issued 1966.)   [*15, 99, 100*]

Papa, K. E. 1971. Growth rate in *Neurospora crassa:* linkage of polygenes. *Genetica* 42:181-86.   [*187*]

Parag, Y. 1968. Phase-microscopic observations of fusions in somatic cells of a heterokaryon of *Schizophyllum commune. Amer. J. Bot.* 55:984-88.
                                                                                                            [*70*]

Parmeter, J. R., Jr., W. C. Snyder, and R. E. Reichle. 1963. Heterokaryosis and variability in plant-pathogenic fungi. *Ann. Rev. Phytopathol.* 1:51-76.
                                                                                                            [*41*]

Parrott, D. M. 1968. Matings between like and unlike populations of potato cyst-nematodes. *Rothamsted Rep.* 1967:147-48.   [*90*]

Parry, K. E., and R. K. S. Wood. 1959a. The adaptation of fungi to fungicides: adaptation to captan. *Ann. Appl. Biol.* 47:1-9.   [*166*]

――――, and ――――. 1959b. The adaptation of fungi to fungicides: adaptation to thiram, ziram, ferbam, nabam and zineb. *Ann. Appl. Biol.* 47:10-16.
                                                                                                            [*166*]

Pateman, J. A., and B. T. O. Lee. 1960. Segregation of polygenes in ordered tetrads. *Heredity* 15:351-61.   [*187*]

Patton, R. F. 1962. Inoculation with *Cronartium ribicola* by bark patch grafting. *Phytopathology* 52:1149-53.   [*63*]

Pelham, J. 1966. Resistance in tomato to Tobacco Mosaic Virus. *Euphytica* 15:258-67.   [*97*]

Pellizzari, E. D., J. Kuć, and E. B. Williams. 1970. The hypersensitive reaction in *Malus* species: changes in the leakage of electrolytes from apple leaves after inoculation with *Venturia inaequalis. Phytopathology* 60:373-76.
                                                                                                            [*124*]

Perkins, D. D. 1962. Preservation of *Neurospora* stock cultures with anhydrous silica gel. *Can. J. Microbiol.* 8:591-94.   [*6*]

Person, C. 1959. Gene-for-gene relationships in host: parasite systems. *Can. J. Bot.* 37:1101-30.   [*107*]

――――, and G. Sidhu. 1971. Genetics of host-parasite interrelationships. *Mutation breeding for disease resistance.* IAEA (Vienna): 31-38.   [*13, 15*]

Pinthus, M. J., Y. Eshel, and Y. Shchori. 1972. Field and vegetable crop mutants with increased resistance to herbicides. *Science* 177:715-16.
                                                                                                            [*171*]

Pitt, D., and C. Coombes. 1969. Release of hydrolytic enzymes from cytoplasmic particles of *Solanum* tuber tissues during infection by tuber-rotting fungi. *J. Gen. Microbiol.* 56:321-29.   [*118*]

Poinar, G. O., Jr. 1971. Use of nematodes for microbial control of insects, Ch. 8, pp. 181-203. In *Microbial control of insects and mites* (H. D. Burges and N. W. Hussey, eds.). New York: Academic.   [*159*]

Pontecorvo, G. 1956. The parasexual cycle in fungi. *Ann. Rev. Microbiol.* 10:393-400.   [*41, 81*]

Powers, H. R., Jr., and W. J. Sando. 1957. Genetics of host-parasite relationship in powdery mildew of wheat. *Phytopathology* 47:453 (abst.).   [*96*]

Priest, D., and R. K. S. Wood. 1961. Strains of *Botrytis allii* resistant to chlorinated nitrobenzenes. *Ann. Appl. Biol.* 49:445-60.   [*166*]

Proverbs, M. D. 1969. Induced sterilization and control of insects. *Ann. Rev. Entomol.* 14:81–102.   [*163*]

———. 1971. Orchard assessment of radiation-sterilized moths for control of *Laspeyresia pomonella* (L.) in British Columbia. In *Application of induced sterility for control of lepidopterous populations.* Panel Proceedings. IAEA (Vienna) 1970: 117–33.   [*159*]

Puhalla, J. E. 1968. Compatibility reactions on solid medium and interstrain inhibition in *Ustilago maydis. Genetics* 60:461–74.   [*74, 75, 79, 86, 87*]

———. 1969. The formation of diploids of *Ustilago maydis* on agar medium. *Phytopathology* 59:1771–72.   [*75*]

———. 1970. Genetic studies of the *b* incompatibility locus of *Ustilago maydis. Genet. Res.* 16:229–32.   [*75*]

———. 1973. Heterokaryosis in *Verticillium:* a different model. *Abst. 2nd. Int. Congr. Plant Pathol.* (Minneapolis):711.   [*61*]

Puranik, S. B., and D. E. Mathre. 1971. Biology and control of ergot on male sterile wheat and barley. *Phytopathology* 61:1075–80.   [*191*]

Pustovojt, V. S. 1965. Science and practice. *Agrobiologia* 1965:3–9 (Russian) (See *Plant Breed. Abst.* 35:4952).   [*97*]

Raa, J. 1968. Natural resistance of apple plants to *Venturia inaequalis.* Ph.D. Thesis. Univ. Utrecht, Neth. 100 pp.   [*124*]

Rahe, J., J. Kuć, C. Chuang, and E. Williams. 1969. Induced resistance in *Phaseolus vulgaris* to bean anthracnose. *Phytopathology* 59:1641–45.

    [*119*]

Raper, J. R., and K. Esser. 1961. Antigenic differences due to the incompatibility factors in *Schizophyllum commune. Zeits. Vererb.* 92:439–44.

    [*150*]

Remington, C. L. 1968. The population genetics of insect introduction. *Ann. Rev. Entomol.* 13:415–26.   [*162*]

Richards, M. 1971. Incidence of viruses in fungi and their potential application. Paper presented at 1st International Mycology Congress, at Exeter, Eng. (Unpublished.)   [*87*]

Riley, C. V. 1872. On the cause of the deterioration in some of our native grape vines, etc. *Amer. Nat.* 6:532–44.   [*99*]

Robbins, W. E. 1972. Hormonal chemicals for invertebrate pest control. pp. 172–96. In *Pest control strategies for the future.* Washington, D.C.: Nat. Acad. Sci.   [*159*]

Roberts, H. F. 1929. *Plant hybridization before Mendel.* Princeton Univ. Press. 374 pp.   [*10*]

Robinow, C. F., and J. Marak. 1966. A fiber apparatus in the nucleus of the yeast cell. *J. Cell Biol.* 29:129–51.   [*37*]

Robinson, R. A. 1969. Disease resistance terminology. *Rev. Appl. Mycol.* 48:593–606.   [*11, 112*]

———. 1971. Vertical resistance. *Rev. Plant Pathol.* 50:233–39.   [*14, 180, 183*]

Roelofs, W. L., E. H. Glass, J. Tette, and A. Comeau. 1970. Sex pheromone trapping for red-banded leaf roller control: theoretical and actual. *J. Econ. Entomol.* 63:1162–67.   [*159*]

Rohde, R. A. 1972. Expression of resistance in plants to nematodes. *Ann. Rev. Phytopathol.* 10:233–52.   [*15*]

Romero, S., and D. C. Erwin. 1969. Variation in pathogenicity among single-oospore cultures of *Phytophthora infestans. Phytopathology* 59:1310–17.
[*48*]

Ross, H., and C. A. Huijsman. 1969. Uber die Resistenz von *Solanum (Tuberarium)*—Arten gegen europäische Rassen des Kartoffelnematoden (*Heterodera rostochiensis* woll.) *Theor. Appl. Genet.* 39:113–22.   [*90*]

Rowell, J. B. 1955. Functional role of compatibility factors and an *in vitro* test for sexual compatibility with haploid lines of *Ustilago zeae. Phytopathology* 45:370–74.   [*75*]

———, and J. E. DeVay. 1954. Genetics of *Ustilago zeae* in relation to basic problems of its pathogenicity. *Phytopathology* 44:356–62.   [*75*]

———, W. Q. Loegering, and H. R. Powers. 1963. Genetic model for physiologic studies of mechanisms governing development of infection type in wheat stem rust. *Phytopathology* 53:932–37.   [*141*]

Sackston, W. E. 1962. Studies on sunflower rust. III. Occurrence, distribution and significance of races of *Puccinia helianthi* Schw. *Can. J. Bot.* 40:1449–58.   [*96*]

Sadasivan, T. S., S. Suryanarayanan, and L. Ramakrishnan. 1965. Influence of temperature on rice blast disease, pp. 163–71. In *The rice blast disease.* Proc. Symp., Int. Rice Res. Inst. Baltimore, Md.: Johns Hopkins Press.
[*82*]

Samaddar, K. R., and R. P. Scheffer. 1968. Effect of the specific toxin in *Helminthosporium victoriae* on host cell membranes. *Plant Physiol.* 43:21–28.   [*113*]

Samborski, D. J. 1963. A mutation in *Puccinia recondita* Rob. ex Desm. f. sp. *tritici* to virulence on Transfer, Chinese Spring X *Aegilops umbellulata* Zhuk. *Can. J. Bot.* 41:475–79.   [*68*]

———. 1969. Leaf rust of wheat in Canada in 1969. *Can. Plant Dis. Surv.* 49:80–82.   [*108*]

———, and P. L. Dyck. 1968. Inheritance of virulence in wheat leaf rust on the standard differential wheat varieties. *Can. J. Genet. Cytol.* 10:24–32.
[*68, 96*]

Sander, E. 1967. Alteration of Fd phage in tobacco leaves. *Virology* 33:121–30.   [*153*]

Sanderson, K. E., and A. M. Srb. 1965. Heterokaryosis and parasexuality in the fungus *Aschochyta imperfecta. Amer. J. Bot.* 52:72–81.   [*42*]

Sanghi, A. K., and N. H. Luig. 1971. Resistance in wheat to formae speciales *tritici* and *secalis* of *Puccinia graminis. Can. J. Genet. Cytol.* 13:119–27.
[*69*]

Sansome, E. 1963. Meiosis in *Pythium debaryanum* Hesse and its significance in the life-history of the Biflagellatae. *Trans. Brit. Mycol. Soc.* 46:63–72.
[*46*]

———. 1970. Selfing as a possible cause of disturbed ratios in *Phytophthora* crosses. *Trans. Brit. Mycol. Soc.* 54:101–07.   [*46*]

————, and C. M. Brazier. 1973. Diploidy and chromosomal structural hybridity in *Phytophthora infestans*. *Nature* 241:344–45.   [*48*]

Sato, N., K. Tomiyama, N. Katsui, and T. Masamune. 1968. Isolation of rishitin from tubers of interspecific potato varieties containing different late-blight resistance genes. *Ann. Phytopathol. Soc. Jap.* 34:140–42.   [*116*]

Savage, E. J., C. W. Clayton, J. H. Hunter, J. A. Brenneman, C. Laviola, and M. E. Gallegly. 1968. Homothallism, heterothallism, and interspecific hybridisation in the genus *Phytophthora*. *Phytopathology* 58:1004–21.

[*45*]

Saxena, K. M. S., and A. L. Hooker. 1968. On the structure of a gene for disease resistance in maize. *Proc. Nat. Acad. Sci.* 61:1300–05.

[*22, 23, 24*]

Schafer, J. F. 1971. Tolerance to plant disease. *Ann. Rev. Phytopathol.* 9:235–52.   [*32*]

Scheffer, R. P., and R. R. Nelson. 1967. Geographical distribution and prevalence of *Helminthosporium victoriae*. *Plant Dis. Rep.* 51:110–11.

[*175*]

————, ————, and A. J. Ullstrup. 1967. Inheritance of toxin production and pathogenicity in *Cochliobolus carbonum* and *Cochliobolus victoriae*. *Phytopathology* 57:1288–91.   [*54*]

————, and O. C. Yoder. 1972. Host-specific toxins and selective toxicity. pp. 251–272. In *Phytotoxins in plant diseases* (R. K. S. Wood, A. Ballio, and A. Graniti, eds.) London, Academic.   [*113, 114*]

Scheifele, G. L., R. R. Nelson, and C. Koons. 1969. Male sterility cytoplasm conditioning susceptibility of resistant inbred lines of maize to yellow leaf blight caused by *Phyllosticta zeae*. *Plant Dis. Rep.* 53:656–59.   [*177*]

Schertz, K. F., and Y. P. Tai. 1969. Inheritance of reaction of *Sorghum bicolor* (L.) Moench to toxin produced by *Periconia circinata* (Mang.) Sacc. *Crop Sci.* 9:621–24.   [*19*]

Schroeder, W. T., and R. Providenti. 1969. Resistance to benomyl in powdery mildew of cucurbits. *Plant Dis. Rep.* 53:271–75.   [*164*]

Schwinghamer, E. A. 1959. The relation between radiation dose and the frequency of mutations for pathogenicity in *Melampsora lini*. *Phytopathology* 49:260–69.   [*68, 141, 142*]

Scott, K. J., and D. J. Maclean. 1969. Culturing of rust fungi. *Ann. Rev. Phytopathol.* 7:123–46.   [*64*]

Seevers, P. M., and J. M. Daly. 1970a. Studies on wheat stem rust resistance controlled at the *Sr6* locus I. The role of phenolic compounds. *Phytopathology* 60:1322–28.   [*127*]

————, and ————. 1970b. Studies on wheat stem rust resistance controlled at the *Sr6* locus II. Peroxidase activities. *Phytopathology* 60:1642–47.

[*127*]

————, ————, and F. F. Catedral. 1971. The role of peroxidase isozymes in resistance to wheat stem rust disease. *Plant Physiol.* 48:353–60.   [*127, 128*]

Sharma, S. K., and R. Prasada. 1969. Production of new races of *Puccinia graminis* var. *tritici* from mixtures of races on wheat seedlings. *Aust. J. Agr. Res.* 20:981–85.   [*68*]

Shaw, D. S., and C. G. Elliott. 1968. Streptomycin resistance and morphological variation in *Phytophthora cactorum*. *J. Gen. Microbiol.* 51:75–84.
[45]

————, and I. A. Khaki. 1971. Genetical evidence for diploidy in *Phytophthora*. *Genet. Res.* 17:165–67.
[47]

Shay, J. R., E. B. Williams, and J. Janick. 1962. Disease resistance in apple and pear. *Proc. Amer. Soc. Hort. Sci.* 80:97–104.
[15]

Shepherd, K. W., and G. M. E. Mayo. 1972. Genes conferring specific plant disease resistance. *Science* 175:375–80.
[25, 134]

Sheridan, J. E. 1971. The incidence and control of mercury-resistant strains of *Pyrenophora avenae* in British and New Zealand seed oats. *N.Z. J. Agr. Res.* 14:469–80.
[164]

Sidhu, G., and C. Person. 1971. Genetic control of virulence in *Ustilago hordei*. II. Segregations for higher levels of virulence. *Can. J. Genet. Cytol.* 13:173–78.
[77, 96, 98]

————, and ————. 1972. Genetic control of virulence in *Ustilago hordei*. III. Identification of genes for host resistance and demonstration of gene-for-gene relations. *Can. J. Genet. Cytol.* 14:209–13.
[96, 98, 99]

Sigurbjörnsson, B., and A. Micke. 1969. Progress in mutation breeding. *Induced mutations in plants.* IAEA (Vienna):673–98.
[16]

Silva, J. 1972. Alleles at the *b* incompatibility locus in Polish and North American populations of *Ustilago maydis* (DC) Corda. *Physiol. Plant Pathol.* 2:333–37.
[75, 80]

Simchen, G., and J. Stamberg. 1969. Fine and coarse controls of genetic recombination. *Nature* 222:329–32.
[39]

Simmonds, N. W. 1960. Variability and utilization of Andean potatoes. *51st Ann. Rep. John Innes Inst.* (Hertford, Eng.):33–34.
[30]

————. 1962. Variability in crop plants, its use and conservation. *Biol. Rev.* 37:422–65.
[190]

Simons, M. D. 1966. Relative tolerance of oat varieties to the crown rust fungus. *Phytopathology* 56:36–40.
[32, 33]

————. 1969. Heritability of crown rust tolerance in oats. *Phytopathology* 59:1329–33.
[32]

————. 1970. Crown rust of oats and grasses. *Amer. Phytopathol. Soc. Monogr. No. 5.* 47 pp.
[18]

————, and H. C. Murphy. 1955. A comparison of certain combinations of oat varieties as crown rust differentials. *USDA Tech. Bull.* 1112:1–22.
[108]

————, F. J. Zillinsky, and N. F. Jensen. 1966. A standardized system of nomenclature for genes governing characters of oats. *USDA ARS Publ.* 34–85. 22 pp.
[18, 22]

Sitterly, W. R. 1972. Breeding for disease resistance in cucurbits. *Ann. Rev. Phytopathol.* 10:471–90.
[15]

Slesinski, R. S., and A. H. Ellingboe. 1971. Transfer of ^{35}S from wheat to the powdery mildew fungus with compatible and incompatible parasite/host genotypes. *Can. J. Bot.* 49:303–10.
[129]

Smith, J. D., W. Arber, and U. Kühnlein. 1972. Host specificity of DNA produced by *Escherichia coli* XIV. The role of nucleotide methylation in *in vivo* B-specific modification. *J. Mol. Biol.* 63:1–8.   [*150*]

Smith, K. M. 1957. *A textbook of plant virus diseases.* 2nd ed., p. 387. Boston: Little, Brown.   [*135*]

Smith, R. H. 1971. Induced conditional lethal mutations for the control of insect populations. *Sterility principle for insect control or eradication.* Proceedings organized by IAEA and FAO (Athens), Sept. 14–18, 1970: 453–65.   [*159, 160*]

Srivastava, B. I. S. 1971. Studies of chromatin from normal and crown gall tumor tissue cultures of tobacco. *Physiol. Plant Pathol.* 1:421–33.   [*151*]

———, and K. C. Chadha. 1970. Liberation of *Agrobacterium tumefaciens* DNA from the crown gall tumor cell DNA by shearing. *Biochem. Biophysiol. Res. Commun.* 40:968–72.   [*151*]

Stairs, G. R. 1972. Pathogenic microorganisms in the regulation of forest insect populations. *Ann. Rev. Entomol.* 17:355–72.   [*159*]

Stakman, E. C., J. J. Parker, and F. J. Piemeisel. 1918. Can biologic forms of stem rust on wheat change rapidly enough to interfere with breeding for rust resistance? *J. Agr. Res.* 14:111–23.   [*10*]

———, F. J. Piemeisel, and M. N. Levine. 1918. Plasticity of biologic forms of *Puccinia graminis.* *J. Agr. Res.* 15:221–50.   [*36*]

Stamberg, J., and Y. Koltin. 1973. The origin of specific incompatibility alleles: a deletion hypothesis. *Amer. Nat.* 107:35–45.   [*144*]

Stephens, S. G. 1961. Resumé of the Symposium. *Mutation and plant breeding.* Nat. Acad. Sci., Nat. Res. Coun. Publ. 891:495–509.   [*16*]

Stevens, N. E. 1948. Disease damage in clonal and self-pollinated crops. *J. Amer. Soc. Agron.* 40:841–44.   [*190*]

Stoner, A. K. 1970. Breeding for insect resistance in vegetables. *HortScience* 5:76–79.   [*15*]

Strobel, G. A. 1973. Biochemical basis of the resistance of sugarcane to eyespot disease. *Proc. Nat. Acad. Sci.* 70:1693–6.   [*113*]

Stroun, M., P. Anker, P. Gahan, A. Rossier, and H. Greppin. 1971. *Agrobacterium tumefaciens* ribonucleic acid synthesis in tomato cells and crown gall induction. *J. Bact.* 106:634–39.   [*151*]

Stubbs, R. W. 1968. *Puccinia striiformis* Westend f.sp. *tritici.* The evolution of the genetic relationship of host and parasite. *Abst. 1st Int. Congr. Plant Pathol.* 196.   [*68, 141*]

Sturhan, D. 1971. Biological races, Ch. 15, Vol. 2 B. M. Zuckerman, W. F. Mai and R. A. Rohde, eds. In *Plant parasitic nematodes.* New York: Academic. pp. 51–71.   [*90*]

Suzuki, H. 1965. Origin of variation in *Piricularia oryzae,* pp. 111–49. In *The rice blast disease.* Proc. Symp. Rice Blast Dis., Int. Rice Res. Inst. Baltimore, Md.: Johns Hopkins Press.   [*82*]

———. 1967. Studies on biologic specialization in *Pyricularia oryzae* cav. Monogr. Tokyo Univ. Agr. Tech. (Tokyo). 235 pp.   [*84*]

Szkolnik, M., and J. D. Gilpatrick. 1969. Apparent resistance of *Venturia inaequalis* to dodine in New York apple orchards. *Plant Dis. Rep.* 53:861–64.   [*164*]

Takahashi, Y. 1965. Genetics of resistance to the rice blast disease, pp. 303–29. In *The rice blast disease*. Proc. Symp. Rice Blast Dis., Int. Rice Res. Inst., 1963. Baltimore, Md.: Johns Hopkins Press. [85]

Taylor, J. 1953. The effect of continual use of certain fungicides on *Physalospora obtusa*. *Phytopathology* 43:268–70. [164]

Temin, H. M., and S. Mizutani. 1970. RNA-dependent DNA polymerase in virions of Rous sarcoma virus. *Nature* 226:1211–13. [153]

Thomas, P. L., and C. Person. 1965. Genetic control of low virulence in *Ustilago*. *Can. J. Genet. Cytol.* 7:583–88. [78]

Thompson, J. M., and J. Taylor. 1971. Genetic susceptibility to *Glomerella* leaf blotch in apple. *J. Hered.* 62:303–06. [192]

Thornton, R. J., and J. R. Johnston. 1971. Rates of spontaneous mitotic recombination in *Saccharomyces cerevisiae*. *Genet. Res.* 18:147–51. [82]

Thurston, H. D. 1971. Relationship of general resistance: late blight of potato. *Phytopathology* 61:620–26. [11, 15, 31]

Timmer, L. W., J. Castro, D. C. Erwin, W. L. Belser, and G. A. Zentmyer. 1970. Genetic evidence for zygotic meiosis in *Phytophthora capsici*. *Amer. J. Bot.* 57:1211–18. [47]

Tinline, R. D. 1962. Cochliobolus sativus. V. Heterokaryosis and parasexuality. *Can. J. Bot.* 40:425–37. [52]

_____, and B. H. MacNeill. 1969. Parasexuality in plant pathogenic fungi. *Ann. Rev. Phytopathol.* 7:147–70. [41, 81]

Tolmsoff, W. J. 1972. Diploidization and heritable gene repression-derepression as major sources for variability in morphology, metabolism, and pathogenicity of *Verticillium* species. *Phytopathology* 62:407–13. [42]

Tomiyama, K. 1966. Double infection by an incompatible race of *Phytophthora infestans* of a potato plant cell which has previously been infected by a compatible race. *Ann. Phytopathol. Soc. Jap.* 32:181–85. [117]

_____. 1967. Further observation on the time requirement for hypersensitive cell death of potatoes infected by *Phytophthora infestans* and its relation to metabolic activity. *Phytopathol. Z.* 58:367–78. [117, 118]

_____. 1971. Cytological and biochemical studies of the hypersensitive reaction of potato cells to *Phytophthora infestans*. *Morphological and biochemical events in plant-parasite interaction* (S. Akai and S. Ouchi, eds.). *Phytopathol. Soc. Jap.* (Tokyo):387–99. [118]

T. Sakuma, N. Ishizaka, N. Sato, N. Katsui, M. Takasugi, and T. Masanome. 1968. A new antifungal substance isolated from resistant potato tuber tissue infected by pathogens. *Phytopathology* 58:115–16. [116]

Toxopeus, H. J. 1956. Reflections on the origin of new physiologic races of *Phytophthora infestans* and the breeding of resistance in potatoes. *Euphytica* 5:221–37. [97]

_____. 1957. On the influence of extra R-genes on the resistance of the potato to the corresponding P-races of *Phytophthora infestans*. *Euphytica* 6:106–10. [21]

Triantaphyllou, A. C. 1971. Ch. 13, Genetics and cytology, pp. 1–34. In *Plant parasitic nematodes*, Vol. 2 (B. M. Zuckerman, W. F. Mai, and R. A. Rohde, eds.). New York: Academic. [91]

Ullstrup, A. J. 1972. The impacts of the Southern corn leaf blight epidemics of 1970–1971. *Ann. Rev. Phytopathol.* 10:37–50.   [*179*]

Vaarama, A. 1947. Experimental studies on the influence of DDT insecticide upon plant mitosis. *Hereditas* 33:191–219.   [*170*]

Van der Plank, J. E. 1963. *Plant diseases: epidemics and control.* New York: Academic. 349 pp.   [*11, 28, 30, 175, 180*]

―――. 1968. *Disease resistance in plants.* New York: Academic. 206 pp.
                    [*15, 29, 175, 181, 183*]

―――. 1971. Stability of resistance to *Phytophthora infestans* in cultivars without R genes. *Potato Res.* 14:263–70.   [*188*]

Van Dijkman, A., and A. Kaars Sijpesteijn. 1971. A biochemical mechanism for the gene-for-gene resistance of tomato to *Cladosporium fulvum. Neth. J. Plant Pathol.* 77:14–24.   [*126*]

Varns, J. L., W. W. Currier, and J. Kuć. 1971. Specificity of rishitin and phytuberin accumulation by potato. *Phytopathology* 61:968–71.   [*116*]

―――, and J. Kuć. 1971. Suppression of rishitin and phytuberin accumulation and hypersensitive response in potato by compatible races of *Phytophthora infestans. Phytopathology* 61:178–81.   [*117*]

Vavilov, N. I. 1949. *The origin, variation, immunity, and breeding of cultivated plants.* Trans. by K. Starr Chester. Waltham, Mass.: Chronica Botanica. 364 pp.   [*12, 93*]

Villareal, R. L., and R. M. Lantican. 1965. The cytoplasmic inheritance of susceptibility to *Helminthosporium* leaf spot in corn. *Phil. Agr.* 49:294–300.   [*34*]

Waddington, C. H. 1961. Genetic assimilation. *Advan. Genet.* 10:257–90.
                    [*32*]

Waggoner, P. E., and J. G. Horsfall. 1969. EPIDEM A simulator of plant disease written for a computer. Conn. Agr. Exp. Sta. Bull. 698:1–80.
                    [*174, 175*]

―――, ―――, and R. J. Lukens. 1972. EPIMAY, a simulator of southern corn leaf blight. Conn. Agr. Exp. Sta. Bull. 729. 88 pp.   [*174, 175*]

Walker, J. C. 1951. Genetics and plant pathology, pp. 527–54. In *Genetics in the 20th century.* (L. C. Dunn, ed.) New York: Macmillan.   [*10*]

―――. 1965. Use of environmental factors in screening for disease resistance. *Ann. Rev. Phytopathol.* 5:197–208.   [*15*]

Wallace, A. T. 1965. Increasing the effectiveness of ionizing radiations in inducing mutations at the vital locus controlling resistance to the fungus *Helminthosporium victoriae* in oats. *The Use of Induced Mutations in Plant Breeding. Radiat. Bot.,* Suppl. 5: 237–50.   [*18*]

Walter, J. M. 1967. Hereditary resistance to disease in tomato. *Ann. Rev. Phytopathol.* 5:131–62.   [*15*]

Wang, C.-S., and J. R. Raper. 1970. Isozyme patterns and sexual morphogenesis in *Schizophyllum. Proc. Nat. Acad. Sci.* 66:882–89.   [*150*]

Ward, S. 1973. Chemotaxis by the nematode *Caenorhabditis elegans:* identification of attractants and analysis of the response by use of mutants. *Proc. Nat. Acad. Sci.* 70:817–21.   [*91*]

Waterhouse, W. L. 1929. Australian rust studies I. *Proc. Linn. Soc. N.S.W.* 54:615–80. [43]

Watson, I. A. 1970a. Changes in virulence and population shifts in plant pathogens. *Ann. Rev. Phytopathol.* 8:209–30. [181, 183]

——. 1970b. The utilization of wild species in the breeding of cultivated crops resistant to plant pathogens, pp. 441–57. In *Genetic resources in plants* (O. H. Frankel and E. Bennett, eds.) Philadelphia: Davis. [13]

——, and N. H. Luig. 1958. Somatic hybridization in *Puccinia graminis* var. *tritici. Proc. Linn. Soc. N.S.W.* 83:190–95. [70]

——, and ——. 1961. Leaf rust on wheat in Australia: A systematic scheme for the classification of strains. *Proc. Linn. Soc. N.S.W.* 86:241–50. [108]

——, and ——. 1962. Asexual intercrosses between somatic recombinants of *Puccinia graminis. Proc. Linn. Soc. N.S.W.* 87:99–104. [67, 68]

——, and ——. 1963. The classification of *Puccinia graminis* var *tritici* in relation to breeding resistant varieties. *Proc. Linn. Soc. N.S.W.* 88:235–58. [108]

——, and ——. 1968a. Progressive increase in virulence in *Puccinia graminis* f. sp. *tritici. Phytopathology* 58:70–73. [68]

——, and ——. 1968b. The ecology and genetics of host-pathogen relationships in wheat rusts in Australia. Proc. 3rd Int. Wheat Genet. Symp. Canberra, Aust. Acad. Sci. (Canberra): 227–38. [72]

Weber, D. J., and J. M. Ogawa. 1965. The mode of action of 2,6-dichloro-4-nitro-aniline in *Rhizopus arrhizus. Phytopathology* 55:159–65. [164]

Webster, J. M. 1965. Controlled mating to produce hybrids between pathotypes of *Heterodera rostochiensis* Woll. *Nematologica* 11:299–300. [89]

——. 1967. The significance of biological races of *Ditylenchus dipsaci* and their hybrids. *Ann. Appl. Biol.* 59:77–83. [90]

Webster, R. K., J. M. Ogawa, and E. Bose. 1970. Tolerance of *Botrytis cinerea* to 2,6—dichloro-4-nitroaniline. *Phytopathology* 60:1489–92. [166]

——, ——, and C. J. Moore. 1968. The occurrence and behaviour of variants of *Rhizopus stolonifer* tolerant to 2,6-dichloro-4-nitroaniline. *Phytopathology* 58:997–1003. [166]

Wehrhan, C. F., and W. Klassen. 1971. Genetic insect control methods involving the release of relatively few laboratory-reared insects. *Can. Entomol.* 103:1387–96. [161]

Wells, H. D., D. K. Bell, and C. A. Jaworski. 1972. Efficacy of *Trichoderma harzianum* as a biocontrol for *Sclerotium rolfsii. Phytopathology* 62:442–47. [168]

Wheeler, H. E., and H. H. Luke. 1955. Mass screening for disease-resistant mutants in oats. *Science* 122:1229. [17]

——, A. Novacky, and H. H. Luke. 1971. Isoenzymes of victoria blight-resistant oat lines selected from susceptible cultivars. *Phytopathology* 61:1147–48. [18]

Whitehouse, H. L. K. 1951. A survey of heterothallism in the Ustilaginales. *Trans. Brit. Mycol. Soc.* 34:340–55. [73, 77]

Whitten, M. J. 1971. An approach to insect control by genetic manipulation of natural populations. *Science* 171:682–84.   [*159, 161*]

Wilde, P. 1961. Ein Beitrag zur Kenntnis der Variabilität von *Phytophthora infestans. Arch. Mikrobiol.* 40:163–95.   [*141*]

Wilkie, D. 1970. Reproduction of mitochondria and chloroplasts, pp. 381–99. In *Organization and control in prokaryotic and eukaryotic cells.* Soc. Exp. Microbiol. Symp. 20. (H. P. Charles and B. C. J. G. Knight, eds.). Cambridge, Mass.: Harvard Univ. Press.   [*42*]

Williams, E. B., and J. Kuć. 1969. Resistance in *Malus* to *Venturia inaequalis. Ann. Rev. Phytopathol.* 7:223–46.   [*15, 107*]

————, and J. R. Shay. 1957. The relationship of genes for pathogenicity and certain other characters in *Venturia inaequalis* (Cke.) Wint. *Genetics* 42:704–11.   [*60, 106*]

Williams, N. D., F. J. Gough, and M. R. Rondon. 1966. Interaction of pathogenicity genes in *Puccinia graminis* f. sp. *tritici* and reaction genes in *Triticum aestivum* ssp. *vulgare* "Marquis" and "Reliance." *Crop Sci.* 6:245–48.   [*96*]

Williams, P. G. 1971. A new perspective of the axenic culture of *Puccinia graminis* f. sp. *tritici* from uredospores. *Phytopathology* 61:994–1002.   [*65*]

————, K. J. Scott, and J. L. Kuhl. 1966. Vegetative growth of *Puccinia graminis* f. sp. *tritici* in vitro. *Phytopathology* 56:1418–19.   [*64*]

Williams, W. 1964. *Genetical principles and plant breeding.* Oxford: Blackwell pp. 504.   [*13*]

Wilson, C. L. 1969. Use of plant pathogens in weed control. *Ann. Rev. Phytopathol.* 7:411–34.   [*173*]

————. 1973. A lysosomal concept for plant pathology. *Ann. Rev. Phytopathol.* 11:247–72.   [*118*]

Wimalajeewa, D. L. S., and J. E. DeVay. 1971. The occurrence and characterization of a common antigen relationship between *Ustilago maydis* and *Zea mays. Physiol. Plant Pathol.* 1:523–35.   [*149, 150*]

Wolfe, M. S. 1972. The genetics of barley mildew. *Rev. Plant Pathol.* 51:507–22   [*57, 139*]

Wood, R. K. S. 1967. *Physiological plant pathology.* Oxford: Blackwell. 570 pp.   [*112*]

Wood, H. A., and R. F. Bozarth. 1973. Heterokaryon transfer of viruslike particles and a cytoplasmically inherited determinant in *Ustilago maydis. Phytopathology* 63:1019–20.   [*80, 87*]

Yamasaki, Y., and H. Niizeki. 1965. Studies on variation of rice blast fungus *Piricularia oryzae* Cav. I. Karyological and genetical studies on variation. Bull. Nat. Inst. Agr. Sci., Ser. D (Japan), 13:231–74.   [*82, 83, 84*]

Yaniv, Z., and R. C. Staples. 1972. A comparison of ribosomes from bean rust and an axenic wheat rust. *Phytopathology* 62:502 (abst.).   [*65*]

Zadoks, J. C. 1961. Yellow rust on wheat: studies in epidemiology and physiologic specialisation. *T. Pl. Ziekten* 67:69–256.   [*96*]

Zentmyer, G. A., and D. C. Erwin. 1970. Development and reproduction of *Phytophthora*. *Phytopathology* 60:1120–27.   [*45*]

Zimmer, D. E., J. F. Schafer, and F. L. Patterson. 1963. Mutations for virulence in *Puccinia coronata*. *Phytopathology* 53:171–76.   [*68*]

Zirkle, C. 1968. An overlooked 18th century contribution to plant breeding and plant selection. *J. Hered.* 59:195–98.   [*10*]

Zohary, D. 1970. Centers of diversity and centers of origin, pp. 33–42. In *Genetic resources in plants* (O. H. Frankel and E. Bennett, eds.). Philadelphia: Davis.   [*12*]

Zucker, M. 1968. Sequential induction of phenylalanine ammonia-lyase and a lyase-inactivating system in potato tuber discs. *Plant Physiol.* 43:365–74.   [*131*]

———. 1971. Induction of phenylalanine ammonia-lyase in *Xanthium* leaf discs. *Plant Physiol.* 47:442–44.   [*131*]

INDEX